Informational Nature of Being

Joy From Deep Within

Informational Nature of Being

Joy From Deep Within

True Nature of Your Quantum Self

——— ᨦᨦᨦ ———

Seeking Eternal Truth and Wisdom Series

Hemant Gupta

To order additional copies of this book, contact:
Xlibris Corporation
1-888-795-4274
www.Xlibris.com
Orders@Xlibris.com
82193

Contents

To My Mother

Informational Nature of Being

ROAD TO DIGITAL DIVINE

BOOK I (Summary)

COMPUTATIONAL NATURE OF MIND AND MATTER

For most of us, our current perspective is deeply rooted in the eighteenth-century science of materialism. The new science of information and quantum computation is bringing a fresh perspective, a new understanding about the true nature of us and our universe. It has profound implications to the way us humans understand ourselves and our universe. Since this newer understanding rests on digital and informational nature of being and has divine-like qualities, I have referred to this as "digital divine." A road to arrive at "digital divinity" has also been my journey to understand this new nature of us and our universe. From this perch, our universe appears like an informational entity rooted deeply in the nature of silence or zero. Broader laws of information seem to describe its nature and behaviors far better than the classical laws of physics. Our physical universe emerges as a computing platform engaged in grand act of quantum and binary information processing. How does one start from a macro view of our perceptual universe and arrive at the computational nature of matter and mind? How do cosmic, emotional, and rational mind arise from this foundation? How does this view impact the concept of my self that I hold deep with in my psyche? The informational and computational description of our universe provides a framework to naturally explain many such difficult questions. As one realizes that this grand informational and quantum computational entity or digital divine is not only rooted in logic, but it is also rooted in love, oneness, or unity consciousness, one embarks on a new understanding of us, our universe, and our divinity; an incredible bridge between science and spirituality. This is an amazing know-how. We can all benefit from this. For more information, please visit author's websites: *http://www.drhemantgupta.com* and *http://www. roadtodigitaldivine.com*.

Acknowledgments

A number of friends and my family members have helped me greatly in writing, editing, and inspiring me to write various contents through their stories. First and foremost I must thank my wife, Seema, and children, Vinay and Monish, for their love, support, help, and inspiration. Besides being my partners in crime of sharing this existence, I must acknowledge that all have been incredible sources of learning for me that has enabled me to understand the nature of my own self and the universe that I share with them. I am truly indebted to the incredible wisdom and love they bring to our existence. I owe deep gratitude to my father, Dr. Braj Kishore, who not only helped throughout with the process of editing but also provided invaluable wisdom through our weekly chats for many years now. I must also thank Raj and Grace for the many, many years of friendship and love. I am also deeply indebted to my younger brother, Shishir, whose marvelous poetry on self inspired me to write this series of books. I am very grateful to my sister Dr. R. Singhal for her support and guidance throughout my life. I would like to acknowledge Al and, Ellen Fletcher, and Teresa for providing me over the years, the meaning of what is being a true saint. I would also like to thank Jona for sharing her wisdom and spirit to live every day with incredible vigor and liveliness. Last but not least, I would like to thank Ruby for allowing me to share her story and inspiring all of us with the courage she has shown and to see what a great heart she has.

⎯ᴧᴧᴧᴧ◦◦ᴇ◔◦◔ᴇ◦◦ᴧᴧᴧ⎯

Chapter 1

Happiness and Joy from Deep Within

Just like you and most, I have struggled with bringing and saving feeling joy and happiness within myself. As I approach fifty years of my life, I find myself to be writing this book about happiness and joy, an activity that is perhaps the least compatible with my training and education. The only qualifications I have to embark on such adventure is that for the last few decades core of my being has been unhappy and dissatisfied. Some of this, perhaps, led to some serious medical conditions such as allergies or hives. I constantly took medications to control the symptoms, felt miserable most of the time, especially in the mornings. I went about my daily work routine throughout this period, sometimes hiding signs of allergic reactions and most of the time sadness, living inside the core of my being with a smile that could fool anyone. My doctors had no remedy for this unhappiness; they were glad to give me antihistamines for life to control my allergy symptoms. In theory, there was no real reason for this state of my unhappiness. I am and have been blessed: a beautiful and loving wife with a good heart, two wonderful boys, a good job, great father, brother, sister, friends, and I could go on and on. Today, I feel joyous from within myself and have been symptom free of chronic allergies for many years. It is not the medication which did it. Although, I must admit the medicines did help relieve symptoms to a degree. In fact, I had to give up medication to heal myself from within. What caused me to be in that poor emotional state and for that long? What turned this around? The most notable change has come from altering my own perspective. It has made an incredible difference in my emotional well-being. In short, my perspective as a whole starting with who I am to how I relate to others around me and how I see the world around me has undergone a deep change. This book is about this change. As I look around, I find my old paradigm widely shared among many from all walks of our society. For most, it leads to conflict and confusion in the form of stress with a wide range of symptoms. Simple unhappiness to diseases such as cancer, heart attacks, diabetes, hypertension, and one can go on are results of stress caused by inculcating such a paradigm. It will be my hope that this book makes a similar difference in your perspective and inspires a happier change in your life. All my life I have been a hard-core man of science. I always looked to science to find answers. Even though my current perspective is not rooted entirely in science, science has been the route to get there. To my readers with little or no scientific background, I have tried to make this message appear as little scientific as my abilities have allowed me to do so. Unfortunately, my inabilities may have rendered moments where I could not explain difficult scientific concepts simply for all to grasp. I offer my sincere apologies to my readers for my subjection. I do recommend readings to clarify these ideas further.

There is so much information on this idea and so many who have taken on to tell so much, yet most of us struggle with being and staying happy. In fact, most of us cannot even imagine our existence where our being is bubbling with feelings of joy and happiness, emanating from our core, for any significant time. Achieving everlasting happiness has been the holy grail of human achievement. The principal promise of religion, spirituality, innovation, education, hard work, or play is achieving immediate or some future state of happiness for self or others. We see all around us people with a wide diversity of perspectives. Many studies reveal religious or spiritual people are happier. Many studies show married people are happier. Many of us believe that childhood is the stage of most happiness. We often wonder if others are happier than we are, especially if we see someone with qualities such as material richness or good looks—handsome, beautiful, or successful or smart; our mind wonders the possibility that they must be happy or at least happier than we must. Most of us can never know for sure how others feel, but in general form our impression that cycles of being happy and unhappy are universal and spare no one. Most of us instinctively deal with these cycles and respond in almost universal way. We try to eliminate or mitigate episodes leading to unhappiness and amplify the states of being happy. We do this to the best of our capacities and expertise; some of us are better able to save joy in comparison to others.

One is sure that feeling joy is important to all of us. Why is that so? One would think the answer should be clear and precise, but it is not. To some, the answer appears to be clear because it "just is." To others, if we cannot be happy again, we believe life is not worth living or, in extreme cases, consider even killing ourselves. Most of us know that chronic unhappiness leads to all variety of illnesses. Even after knowing this so clearly, most of us end in unhappy states, most of the time. Latest scientific studies have repeatedly shown that the human brain is wired for bliss. Still happiness eludes most of us. What does make us unhappy? What does keep us unhappy?

Does happiness or joy provide satisfying final meaning to our existence? Our journey will begin by addressing the core issue of meaning. Does meaningful existence make us happy or joyous? Or joyous existence is meaningful in itself. This immediately leads to an age-old question, who decides the meaning? Philosophers have struggled with this question for decades. Or "who is it that needs to be happy or decides meaning?" Of course, since we are talking about "our" happiness or happiness of "self," we naturally conclude that it is self or "I" that is at the root of all this inquiry. We end up asking "who am I?" Alternatively, "Who is this self?" Once we understand the true nature of "I," we can address the happiness and meaning associated with "I."

As we search for the "I" that likes to be happy, we find several candidates. Of course we start with the most obvious, me with my name, titles, education, place in the society with certain face, body weight. Myself, the one I can see in the mirror. Great! Are we getting close, or it just seems that way? Then we may be awakened by that voice, "I think I know that this is not all I am, maybe there is more to myself than my physical self." Who is doing the thinking? Who is it that is doing the act of understanding or doubting? Maybe there is a little "I" within my physical body that I cannot see in the mirror. Is this the one which needs to be happy? Has science found any such entity called inner I? It is clear that most of us do recognize the big I or the physical I. However, in a quiet and unobvious way, we all seem to recognize this other "I" or inner "I," especially as we express ourselves in our everyday language and describe our feelings. Here is a real case of a mother who described the following:

Expression of Self

"I'm in my early thirties and the married mother of two young children. I have a good job, and my husband and I get along well. My problem lies within myself. I suffer from something I can only describe as "self-loathing." It started as a teenager (with cutting my arms, drinking, smoking, running with the wrong people). Now I try to keep it all neatly tucked away in my psyche. I've been to therapists and take antidepressants, but this lingering self-hate always surfaces. My symptoms cause me to withdraw, hit myself with hangers, and say and think the most horrible thoughts about myself. Even with my accomplishments, I do not think much of myself. I'm not suicidal, but I often entertain thoughts of cutting my arms and legs or having someone else beat me until I'm black-and-blue as though I deserve punishment for being who I am. I compare myself to others nonstop and sometimes withdraw for days if I meet someone I envy. It's awful! In addition to antidepressants, I've resorted to taking the painkiller Tramadol daily as it tends to lift my mood and help with these feelings of inadequacy. I do not want to pass this on to my kids, whom I love more than anything. Why in the world won't this stop?"

It is clear that an inner "I" exists, at least in our expression, and is intimately entangled with our concept of self. It turns out that understanding nature of this inner "I" is essential for our happiness. As I will describe later that it is the inner "I" that needs to be happy and keeps us contented and joyful. Where does this inner "I" reside? It is a question which humans have been searching for since ancient times. Could it be any organ in my body or a cell or a molecule or may be an atom that is part of my being? Or perhaps a combination of many atoms or molecules or cells or maybe organs that represent this inner "I"? Or on the other hand, does this inner "I" exist beyond space-time, in another domain? One thing is clear, "I" that seemed so simple at the start is not so. Between the physical "I" that I can see in mirror and the informational inner "I," I am confused. Which is my true self? Our journey will begin, by understanding the nature of "I." Then we can address what makes it happy and what will keep it happy.

Approaching the subject of this inner "I" has not been easy, and it has been a goal of humanity since knowledge began. A poet elegantly provides a fitting description of the difficulties associated with understanding the " in the poem entitled "I Know Not Who I Am".

Perspective, My View of Self and Whole

A perspective related to my view of who I am and how I see myself to be connected to all that surrounds me is the most notable mark of human existence. It affects all of our lives. It provides us meaning or may be the meaning of meaning. Is there something I can change about my perspective that can intensely affect my pleasure meter? The answer is yes. One can become blissful in an instant by effectively changing one's perspective and understanding toward one's life, existence, reality, and eternal truth.

I Know Not Who I Am

Shishir Gupta

I know not who I am
I need not know who I am,
Nobody knows who I am,
I will never know who I am,
You need not know who I am,
For that knowledge is best not obtained,
As it has so much to reveal,
That you would be required to know,
All that is around you,
And with all that you interact,
In time and space,
And in realm of ideas and concepts,
That it will boggle your brain,
For a lifetime and more,
So it is best that,
You continue your journey,
Quite unknown and unnamed.

It is my perspective of the nature of my true self and the nature of its perceived reality. To most, the nature of reality as understood by our collective conscience still rests on our perceptual beliefs, conventional deterministic ideas based on Newtonian physics, which has deeply penetrated all aspects of modern human soul and psyche. The legacy of this science is the proposal and reinforcement of self as distinct and separate entity. I am me and you are "you," and we are all separate. We have believed in such proposal for many generations as our senses reinforce such understanding. The science of physics has undergone extraordinary development. The newer findings are radical in nature and are nonintuitive. They present a different picture of the reality that we belong to and offer a new image of ourselves. The new image is such that "I" is not so isolated from the rest. In fact, the rest is entangled with the "I" in a manner that cannot be easily separated. This view presents a different model of behavior. For example, hurting others may mean that you are hurting yourself. Stealing, giving, all have new meanings. Helping others is like helping us, and therefore, it feels so good as well.

With advancements in science of quantum physics, the role of observer has become central to grasping the nature of our true reality. We will start our journey by addressing this profound issue. Human observer and the means or machinery to observe are critical to make sense of the observed universe. Is it the true nature of our reality? Our senses confirm that we are separate from the rest. Nevertheless, the connectedness at the quantum level as one defies this description of reality. How do limitations of our senses affect our view of the true nature of us and the reality we observe and interpret? Can one address the nature of unconscious matter or "I" less matter, conscious matter or matter with "I," such as us or maybe life in general and arrive at the possibility of all to be intimately connected to nonlocal or grand cosmic mind? What kind of perspective leads us to unity consciousness as implied by science of quantum mechanics? Does the quantum nature of universe turn our concept of reality upside down?

SELF or SIMPLY INFORMATIONAL SELF

As I described in Road to Digital Divine, the first book of this series, our modern understanding of our universe is rooted in science of information and computation rather than the old laws based on classical materialism. The new broader laws of information are superseding old classical laws of matter and energy. Our universe emerges as one informational entity or a grand cosmic mind or digital divine where all entities are reducible to entangled quantum bits that are instantaneously connected to one another. It is as if our universe is a grand living entity made up of digital information bits or quantum bits, the informational atoms of this new universe. It implies that we are all informational entities as well and are all digitally or informationally connected to this grand entity. The consciousness from this perch will be viewed as intelligence that separates us from this grand informational whole.

A simplified informational image of "I" or myself emerges. It arises by the interaction of three minds. First, the cosmic mind, which contains all that is material and immaterial connected though local and non local interactions. Second the emotional mind, which is my physical body. It is contained in the cosmic mind. Third the rational mind, which resides in my emotional mind. In other words, "I" or self is an informational entity, comprising of the systems of devices guided by two local minds or intelligences, viz. rational and emotional but rooted in grand quantum information distribution system or the cosmic mind.

My self, with the help of these three minds converts randomness into information allowing me to operate as an informational entity and make sense of my world. What is its true nature? Is there a connection of this informational "I" or self to the inner "I" we talked about earlier? Does knowing our true nature make a difference in our emotional state of being? The answer to both these questions is "Yes". We will find that our happiness very much rests on our understanding of the true nature of "I" or self which is ultimately rooted in broader laws of information.

The informational self has two basic natures arising from its roots in laws of information. First, "I, mine, or my," emanating from the emphasis on binary computation leading to evolution of egoic entity, a widely recognizable human personality trait in modern societies. I, my, and mine signify the focus of self on self-interests or goals to benefit ourselves rather than others. Second, "us, ours, and we" emphasizes the nature of this self that is for others. I propose that this nature of self arises from the basic nature of the quantum BIT itself leading to altruistic behavior found wide spread in the animal kingdom. Therefore, the self that arises from this emphasis is what

I have described as our quantum self. It acts for the benefit of community rather than the individual self. It gives rise to saint like nature in humans. It will lead to a very different community than the community full of egoic entities. Which community will ultimately prevail? Which of these two extreme tendencies of self will survive? Which of the self is my true nature? The knowledge of this truth is not trivial, and in fact, it is nothing short of enlightenment. Once understood mentally and experientially, it has potential to fill us with the incredible joy emanating from deep within.

———~~⋆⊙⋆⊙⋆~~———

Nature of Self

Chapter 2

Nature of Reality
Observer
and
the Means to Observe

Nature of Reality

Classical physicists paid meticulous attention to their sensual impressions. They felt secure that what they sensed was true from the microworld of small and invisible to the macroworld of large and visible. It was "what you see is what you get." The arrival of quantum theory, however, has put a strain on the credibility of our senses. Considering all the ways everything is connected, at the quantum level, the informational entity the whole or the cosmic mind as described in the book 1 of this series Road to Digital Divine defines a new reality of everything in the universe as one, through local as well as nonlocal interactions. If everything is one, how do we understand the world as made up of many? It certainly defies our normal view which is "I am separate from others all the way from physical appearance to bank account." Clearly, the question of the true nature of reality is not as simple as many classical physicists assumed. Is there another reality that transcends our sensual views? If it is, we need to understand what is it, where it is, and how to access it?

Is reality observer dependent? As difficult to believe it is, the science of quantum physics is leading us to believe that, in fact, is the case. According to this science, any notion of reality requires an observer to observe the reality or document what is observed as a first step. It then requires an information processor or a mind to make sense of what can be inferred from what is observed as reality. For simplicity, I will refer to the observer as self. We humans, of course, are among many observers of our universe. We have been a rather keen observer of our universe. We observe our universe with rather developed rational mind and a strong sense of separation. We may, in fact, have the dubious distinction of not only being an observer with abilities to see the whole but even judge and doubt it as well. In today's world, our self has given rise to a reality that incorporates true sense of separateness in its formulation. Our local minds, both emotional and rational, have played a critical role in constructing such reality. Our current understanding leads us to grasp reality from the following two points of views:

1. Abstract, Mathematical, Computational Reality
2. Physical Reality

Both these realities exist for us primarily because of the construct of our local minds. As we saw earlier, in book 1 of this series, Road to Digital Divine, both these realities also exist in the cosmic mind. The cosmic mind contains this physical universe, which arises from deep mathematical or computational abstract reality living at its core. Of course, our local minds are contained in the cosmic mind. The observer's mental reality mirrors the physical reality, chiefly dependent on the hardware and software it has. It construes the reality from its perception, calibration, and analysis. The two realities, physical and mathematical, are eventually connected at a deep level, into one reality in an observer's mind, just as they are connected in cosmic mind. This level will be truth where all becomes one. That is, observer, observed, and observation merge in one wholesome reality.

Mind-Based Physical Reality, Nature of Our Perceptual Reality

The role of the observer in understanding the nature of reality has become clear over the last century or so. It has been one of the most debated areas of physics over last many decades. The Wysiwyg, short for "what you see is what you get" hypothesis is based on two simple postulates: first, experience is the only reality for an observer; second, there is no reality other than experience. Mihai Drăgănescu, 1999, describes the existence of many levels of information processing taking place in our brain:

a. The highest level or psychological level: It may be seen as macroscopic reality or top-down reality comprising of behavior, intellectual activities such as thinking, sentiments, will and desire, et cetera.
b. The neuronal level: It comprises of the networks of neurons, modules, and other structural organization of the brain.
c. The molecular level: Comprising the molecular activities inside and outside the cells for example informational molecules.
d. The quantum level or the bottom-up view as all above rooted in this reality.
e. The experiential level: Or a fact, qualia, experience of the brain and mind reality or result of the computation.

He places experience as the fundamental reality of nature. It puts observer at the center of accessing this reality and its experience as the root of the reality itself. Are qualia or feelings or experiences made of a different "stuff" than the sum total of experiences at the highest level, the neuronal level, the molecular level, and the quantum level? In fact, many have argued that experiential reality is all there is. Then, there is this hypothesis of "now." Is the experience of "now" or present completely and totally defines reality? Is present experience all there is to the reality of our existences?

If visible universe represents a physical and mental act of computation, experience represents the result of these computations. Just like in living systems, physical, mental act of computation, and experiential act make the whole; the analogy applies to universe where a cosmic mind and physical visible universe make a complete whole. From this point of view, the physical act is the hardware and software of our existence, and experiential act is the result of such an act. We will come back to the basic essence of this experiencial reality later in our discussion. First let us examine the nature of calibration and information processing that leads to such experiences.

Humans have been endowed with five senses, which have helped us, not only to survive but also to make sense of our existence. Classical physicists assumed that reality is what we see through human senses. Other life-forms also compose reality based on their sensual inputs. Is reality, framed by humans, very different from the sensual view of other life-forms? Are we dependent on the hardware and software of our senses to make sense of our universe? The answer is yes, we are, in fact, very much dependent on both our hardware and software to make sense of our reality!

Making Sense of Senses

Survival for all life is a dangerous game. Evolutionary biologists name this game as "survival of the fittest." There are so many ways a living being can die. It is incredibly dependent on its surroundings to survive. Some things are useful; some are dangerous and harmful. Life-forms must know how to distinguish between the two. Clearly, knowing the present, past, and future environment may have an incredible impact on winning or losing this game of survival.

A single cell can have as many as hundred thousand chemical receptors on its cell membrane providing it information about the external environment.

For more complex life-forms, sensory apparatuses have evolved that convert inputs from the external environment into an understanding that allows the life-form to have a map of the external world that enable it to survive. Different life-forms display surprisingly wide number of adaptations to their sensory apparatuses to make sense of their environments.

Some of these go far beyond the detection abilities of humans. For example, ants can detect even a few centimeter movement of the earth. Cockroaches are skilled at detecting movements as small as few thousand times the diameter of a hydrogen atom. Crab has hairs on its body including claws to detect tiny movement caused by water currents and vibration.

While we humans use visible light or sight and sound waves or hearing to locate objects, dolphins use echolocation primarily based on high frequency sound waves for detecting movement and finding objects. These sound waves are reflected from objects in the form of echo, which is interpreted by dolphins. Dolphins can hear sound waves with frequencies as high as 100,000 hertz (Hz). Bats can find food up to eighteen feet away and gets information about the insect using their sense of echolocation based on similar principles. Cats have hearing that utilizes audio frequencies that range between 100 and 60,000 Hz. Mice also have abilities to hear high frequencies ranging between 1,000 and 100,000 Hz. Pigeons, on the other hand, can detect low frequency sound waves with frequencies as low as 0.1 Hz. Crickets can hear using their legs. How is that possible? It is because a thin membrane on the cricket's front legs vibrates as it encounters sound waves. Dogs have olfactory membrane up to 150 square centimeters which enables them to hear sound wave with frequencies as high as 40,000 Hz. By comparison, we humans can only hear frequencies between 20 and 20,000 Hz. Despite such limitations, we have produced the best singers and rock band among the whole living kingdom; at least that is what we think!

If you have seen a king cobra dancing on the musical beat of a flute, you may conclude that at least snakes may enjoy our human music. Unfortunately, snakes have no external ears. Therefore, contrary to popular belief, they cannot hear the music played by a snake charmer. They instead are responding to the movements of the snake charmer and the flute.

We humans consider ourselves as the connoisseur of yummy food. With a dubious distinction of being perhaps the only species with the largest number of gourmet recipe for our food, we are proud of our pallet. Pig's tongue contains fifteen thousand taste buds, which, in comparison with the human tongue, is rich in accessing taste as human tongue only has nine thousand taste buds. Bees have taste receptors on their jaws, forelimbs, and antennae. Butterflies also have taste receptors on their feet. Entire body of earthworm is covered with chemo receptors, or one can say taste receptors. Blowflies taste food or external environment using three thousand sensory hairs on their feet. The tongue of snakes as a stark contrast has no taste buds. Instead, the tongue brings smells into the mouth.

Worker honeybees have a ring of iron oxide in their abdomens. This particular magnetic oxide gives them ability to detect changes in the earth's magnetic field, and honeybees use this information for navigation purposes. Bats, endowed with infra red receptors, can detect heat of an animal as far as 16 cm away. Shark has specialized electrosensing receptors with thresholds as low as 0.005 microvolt per centimeter. These receptors may be used to find prey.

Normal vision for humans is twenty-twenty. A hawk's vision equals 20/5. This means, what most humans can see from a distance of five feet, the hawk can see from a distance of twenty feet. A falcon can see a 10 centimeter large object from a distance of 1.5 km. While we have one precious lens per eye, a dragonfly eye contains thirty thousand lenses. The eyes of the chameleon can move independently and observe two directions at the same time. Wouldn't that be neat if we could turn that trick?

Penguin has a clear vision underwater. They are also able to see in the UV or ultraviolet range of the electromagnetic spectrum. Pigeon with eyes mounted laterally on their heads can view 340 degrees, everywhere except behind their heads. The ants can see polarized light. Bees can also see polarized light. In addition, bees can see the light between wavelengths 300 nm and 650 nm.

When we see blurred images, we need to wear glasses or have corrective surgery to be able to see clearly. Nocturnal animals naturally see mostly crude shapes or just outlines. They see no colors. Their sensitivity to lowlight levels is heightened with various adaptations. It is usually enough for them to survive in the dark. The size of their eyes is usually large, with a wider pupil, larger lens, and increased retinal surface to collect more light. Some animal species have adaptations, such as tubular eyes, which increase their size, again to receive more light. Some nocturnal animals, for example, owls have devised extraordinary rotational capacity in the neck. They can rotate their neck through 270 degree, and this, of course, aids their vision.

Clearly senses have evolved to aid living beings in understanding their environment. In fact, living strategies of most life-forms have centered on the sensory apparatuses they have been endowed with. Does input from sensory apparatuses reflect the true nature of reality? The answer lies in calibration of the sensory apparatuses. Moreover, this calibration is often not geared toward sending a true reality to the organism but often to aid survival.

Just like other life-forms, we humans rely on our senses as well as our mental calibration to make sense of our environment. We often forget this calibration aspect of our senses and often get trapped in a reality that is perceptual. Our vision, for example, paints a picture of the visible universe to us. The human eye can process image at the speed of roughly twenty-four frames per second. At this speed, we begin to see connected images even though in reality these images may be still and not connected. This is called flicker fusion rate. Human's eye has a flicker fusion rate of only twenty-four frames per second in dim light, which increases to sixty frames per second in bright light. This is the basis of several trillion of dollars worth multimedia industry flourishing in our modern human society. Eye of a fly is actually superior to us in that regards. It has much higher flicker fusion rate of three hundred frames per second. Imagine, what breed of multimedia industry could flourish in their colonies?

What is displayed on a movie screen, is simply pixels turning on and off. It is our mind or calibration of senses that makes sense of these sequences of pixels turning on and off and converts it into a comprehensible story form. Many times, we know the true reality is different but choose to indulge in such artifacts created due to our sensory opinions. Why? Because it is fun, experiential fun. Our conscious mind has kept up with times and figured out that such reality is nonreal, but our emotional mind still behind times and releases molecules of emotions as if the sensory view resulted from a real incident. Otherwise, how would it be possible to have billions of dollar worth of pornographic industry when humans know for sure that pornographic images they see are not real? What calibrations of our mind would lead to such a behavior pattern? So many of us indulge in romance, love story, or drama when we watch television or movie or while reading a book, even though we know that it is just a story and may not have an ounce of reality associated with it. Why? Because we enjoy every bit of emotional rigmarole it creates within our being.

One of the most favorite activities of my two sons is to visit Universal Studios, which is located close to our home in Los Angeles. Monish, my younger son, enjoys the "terminator" ride while my older one Vinay's favorite has been the "back to future" ride. Visiting Universal Studios is a great place to experience one such pseudo reality. Kids and adults alike love the experience. The rides like "back to the future" and the "terminator" provide audiovisual 3-D feel that almost fools our brain in believing an alternate reality even though we know the experienced reality is not real. For example, in the back to the future ride, the visitors simply are sitting on a seat that undergoes movements, which are coordinated with the projected images that are displayed to the viewers. A combined experience takes the viewers through time travel into the future, with caution as Biff Tannen, a 1955 graduate of the Hill Valley High School has escaped his time. He is now running amuck in the space-time continuum. The doors of the time machine close, Dr. Brown uses his remote control to control the time machine, hovers, and speeds up to 88 miles per hour. With sparks coming from the time machine, it speeds through the open door, blasting through the space time vortex. And the ride begins.

First, Biff leads us to Hill Valley in 2015, where we chased him through town. We smashed into bright neon lighted signs, flying over the town as the chase culminated at the iconic clock tower. He then left for the ice age. We followed him, and slowly, lowered our time machine into the ice age. Biff caused an avalanche that damaged our car. Flying out of the icy caverns, we saw Biff shoot away into time, but our own engine failed and began to plummet down a waterfall. Dr. Brown manages to restart the car, speeding up backward and through time into the Cretaceous Period.

We followed Biff's car into a dormant volcano where there is a tyrannosaurus. Tannen goads it into attacking us, we barely escaped. The dinosaur strikes Tannen's car. It sends it flying, out of control. The dinosaur then swallows our car, only to spit it out few seconds later. Biff's DeLorean is damaged and unable to maneuver, moving down rapidly in an active lava flow toward the edge of a cliff. As our car speeds up to time travel speed, it bumps Tannen's car, sending both of them back to the starting point of the ride. Biff gets out, thanks us and Dr. Brown, but is soon nabbed by security and taken away.

As the ride ended, it was obvious that my older one understood that it was a sensual artifact, and there was no Biff chasing and tyrannosaurus posing imminent danger. My six-year-old, Monish, was a different story. It appeared that he believed in his experience of the ride. As the lights turned on, I tried to explain him that he went nowhere but was sitting on his seat the whole time. When he expressed his disbelief, I let him believe in the reality that his experiences had created for him. There was no hurry for him to know my version of reality. Who knows, it may be no better than the reality that occupied his mind. Is it possible that most of what we experience is nothing but an artifact of our sensory opinion? Or is there a deeper reality?

Mind-based Abstract Mathematical Reality, Nature of a Deeper Reality

Our senses make us aware of the physical reality. If we cannot trust our senses, then what is it that we can trust? Is there a deeper reality that is rooted in our mental analysis? Our mental reality is actually an attempt to comprehend the deeper meaning behind the viewed physical reality. It is the domain of our mind to model a reality from the data from senses. Of course, to be able to do so, mind must be equipped with appropriate hardware and software. Our scientific modeling of such observations is based on abstract mathematics. Mathematics, just like poetry, is a language to express processes within our mind. It has turned out to be so useful in predicting physical reality that many believe that mathematical reality is perfection or true reality while physical reality is a display of this perfection. From this perspective, mental reality may be at the boundary of the mathematical reality and the observed physical reality.

The mathematics of quantum physics uses complex analysis and complex numbers with imaginary and real components. A question immediately emerges. How do complex numbers relate to physical reality? According to Leibnitz, "The divine spirit found a sublime outlet in that wonder of analysis, that portent of the ideal world, that amphibian between being and non-being, which we call the imaginary root of negative unity." Descartes rejected complex roots and coined the derogatory term imaginary to describe complex numbers. Perhaps the easiest way to think of "complex numbers" is as a means of bookkeeping two different values at a time for those physical quantities that do have two different aspects to their nature. Complex numbers provide computational assistance, especially in areas related to applied mathematics such as electromagnetism, quantum mechanics, or fluid dynamics. The computational tools such as contour integration provided by the theory of complex numbers lead to a concept that involves integrating a function over a curve in the complex plane rather than just the curve in the real plane.

Complex numbers are extensively used in physics to simplify calculations. The underlying principles still deals in real quantities, and the complex numbers are "just" a mathematical convenience. In quantum mechanics, the wave-function is in general complex. It is, however, analyzed so it makes predictions in real quantities that can be observed in experiments. For example, the function "probability density," which is real, is derived by the squaring the absolute value of a wave function.

Mathematical Creatures Living in the Depth of Emptiness

We learn from evaluating quantum mechanics that the physical universe obeys the mathematics of quantum mechanics everywhere it has been tested and can be tested. So we obtain the following two conclusions. First, all physical existence including our body mind is solely and entirely governed by mathematics of quantum mechanics.

Second, we can deduce, with high confidence, that there is no other form of matter other than the wave function. Thus, the "true nature" of physical or nonphysical existence consists of the wave function and the wave function alone. Are these the creatures lurking in the depth of emptiness or unmanifest reality?

In the classical Newtonian scheme of physics, the reality and the future were deterministic. Once the universe was set in motion, its fate was determined for all time. There was no freedom and no free will. The mathematical creatures of quantum physics do not join us into a deterministic future. They have free will with endless possibilities of multiple futures, much like our observed reality!

However, our experience of the macroscopic physical world does not follow from the realm of these quantum creatures. In fact, it provides a constant test for our rational mind to make sense of a reality that could be consistent with inferred reality of these quantum entities. What would be the nature of our universe if macroscopic objects behaved like these mathematical creatures? At least, some difficulties that underlie such challenge become clear when we ask such a question.

Our impression of measuring the property of an object at the macroscopic levels, such as the position of a rock, yields a single value for any given state of the object. This is a fundamental belief that allows our rational mind to perform computations that enable actions, most of which we take for granted, such as to catch a ball, avoid bumping into other objects, drive a car, etc. In the world of electrons and other subatomic mathematical creatures, this basic assumption does not hold true as measurement can yield more than one value challenging the calibrations of our rational mind at the most basic level.

If all objects in our macroscopic world behaved like these quantum creatures, our rational mind will be confused, and we will not be able to perform any of these activities, at least not as easily as we do now.

For example, in the quantum world, a single particle can be in many places or states at once! These different states are mathematically described as "superposition states." Once established in these states, the particles act as one entity, irrespective of distance. In other words, two or more particles may be separated by vast distances and yet can act as a single entity. Our rational mind cannot make sense out of such oddity, which is common in the world of these mathematical creatures. Einstein's rational mind did not take this lightly. He referred to this as "spooky action at a distance" and perhaps correctly.

Let us look at another example of measuring a property of an electron called "spin." One can only observe one of two values, "up" or "down." If one measured spin of many identically prepared electrons, half the time one will observe the value "up," and the other half the time one will observe the value "down." We will gain one or the other value, but which value we gain is apparently determined by the toss of a coin. This is exceedingly unusual for our rational mind. It expects a certain value which is unaffected by the act of observation.

Before a measurement, the electron exists in a state of being where both spin up and spin down exist simultaneously, a superposition spin state. At the macroscopic level, there is no state of existence corresponding to this state of quantum matter. This is actually an incomprehensible state of being for a human rational mind. There is no calibration of our binary rational mind to comprehend such a concept. The closest a human rational mind would come to describe it would be a state of "nothingness" or "zero" or "emptiness."

From the mathematics of quantum physics, we know the state of superposition is a state of one mind or community mind. It represents quantum computations. Due to extraordinary quantum parallelism, each of the qubit looses its individuality and participates in calculations as one cosmic mind or entity. Our mathematical creature, electron, resides in this state before being observed. Let us try to understand this further with a simple example.

Let us say you, John, and Jane are given $1,200 to distribute among the three of you. Your share of the loot is $400. You honestly do not need money, but who says no to money. Why not keep it and find some use later on? Jane has not paid rent and needs $1,200 to cover this month of rent. Since she has lost her job, she has no other way of getting this money. John also does not need money, but he plans to keep it for future use as well. You are being generous to John and have promised him $100 from your share of the loot. Nevertheless, you genuinely do not appreciate Jane (past issues) and have wowed never to help her. You figure that after the transaction is complete, you should have $300, Jane should have $400, and John should have $500.

Now imagine, through some divine intervention, all three of you were shrunk down to three quantum particles and thrown into a superposition state. Since no one has seen that state, no one can actually tell how three of you looked like as one.

But when you, through another divine spell, are brought back to your large classical state, and you happen to check your pocket, you would find it empty. John also would have nothing from the original loot. Jane would have all the $1,200. Divine has spoken. As one entity, one may speculate, individual self disappears, and only the need of the whole are paramount. Since Jane genuinely needed the money most, she got it.

The key take away from the above story is the oneness of the superposition state. No matter how many entities are thrown in the mix, they all become one. This supremely fundamental expression of the unknown nature of the quantum superimposed state until a measurement is made flies in the face of all classical experience and lies at the core of quantum mechanics, as we know it today.

Getting back to our example of electron with spin states, since best our rational mind can do is only to view electron in either spin up or spin down, that too after an observation, it can only speculate about this superpositional state of its being. In order to turn this abstract informational state of superposition into a real physical value of the spin, another process must take place. This is the process of observation itself. It is known as "collapsing" the wave function into one of its particular values that the measured spin state is expected to occupy.

Let us understand this feature of quantum physics with an example of a macroscopic object behaving like a quantum entity or our electron. Suppose that a rose represents the electron and instead of spin up or down property, it changes color. It can appear either red or yellow. One-half of the times we look at this rose it appears red, and the other half of the time, it appears yellow. It will be quite a funny sight for a human rational mind to see a rose that sometimes is red or sometimes is yellow. Hmm . . . weird. So what color does the rose have when I am not looking at it? Is it both red and yellow? Is it orange? On the other hand, does it even have a color at all? Or it is a rose even? Alternatively, is the question even meaningful? This is the reason why most scientists call the quantum universe as weird.

One will be obvious to a human rational observer that something even weirder is going on. Whatever the color of the rose, it arises out of the mysterious state of superposition. In fact, not only rose, all that is observed in the physical world arises out of this state of superposition or nothingness. The act of observation itself creates matter or all that is observed.

So what happens when we do look at the rose? Just like a science fiction movie, rose creates itself with lightening speed, or even faster, from nothing to a beautiful rose red, no yellow or may be red! Except that, it is not science fiction.

It is true and exactly what quantum physics implies. Only if Gordon Gekko learned about this process and the speed at which it takes place, he would have found his formula to make billions if not trillions of dollars. Greed is good especially when you can create something from absolutely nothing. Considering the price of gold now days, I dare to venture that his choice of that something out of nothing would not be rose; it would be that big golden nugget!

Clearly, at the macroscopic level, this process of creating something out of nothing could wreck havoc with our perceptual reality. That is contrary to everyday experience where objects occur at one place at one time. Either they are here, or they are not, and that does not depend on whether we look at them. In 1935, Erwin Schrödinger described a thought experiment encapsulating the uneasy coexistence of the quantum and the everyday macroworlds. This is the famous Schrödinger's cat experiment. It goes like this.

A cat is placed in a sealed box. Also, in the box is a device that contains a small amount of a radioactive element linked to poison containing flask. There is a 50 percent chance that one of the unstable atoms will decay within the observation period. If an atom decays, the device breaks, open a poison-containing flask, killing the cat. If there is no decay, the cat lives. Is the cat alive or dead? Can one know which is true without looking?

The cat, according to principles of quantum physics, will stay in both states alive and dead and only when the box is opened and observation is made, is its outcome will be sealed. Obviously, that is bizarre. How can something alive be both alive and dead at the same time? One would tend to think that such a state can only reflect only the weirdness of quantum particles. But that is not the case!

Instead of a cat, a team led by Dr. James E. Lukens and Dr. Jonathan R. Friedman, physicists at the State University of New York at Stony Brook, used a small square loop of superconducting wire. The group was able to put two possibilities, related to the direction of the flow of electric current: clockwise around the loop or counterclockwise into quantum superposition.

As they measured the energy difference between the two states of the loop, they were surprised that resultant current reflected a quantum superposition and not simply flipping directions. Just as the cat is neither alive nor dead, the current flows neither clockwise nor counterclockwise, but is a combination of both the possibilities.

Earlier quantum superposition involving an electron or an atom or even a buckeye ball made of sixty carbon atoms was demonstrated. These researchers showed, for the first time, that the flow of billions of electrons can also be forced into a quantum mechanical state of superposition.

"We observe our car to be in one parking space and not the other, and for people to be on one side of the wall or the other," Dr. Friedman said. "But this says, at least on the scale of a hair-width size superconducting loop, we can have two of these macroscopically well-defined states at the same time. Which is something of an affront to our classical intuitions about the world?"

This work, reported in the July 2000 issue of the journal Nature, was "the first one in which there has been reasonably foolproof evidence you do have a superposition of macroscopic quantum states," said Dr. Anthony J. Leggett, a professor of physics at the University of Illinois at Urbana-Champaign, who first proposed the experiment in 1985.

Could the timescales describing the observation give us clues to understand the underlying processes? Clearly the processes are fast and time scales exceedingly small. Ferenc Kraus has been able to record the shortest time intervals ever viewed, in his lab at the Max Planck Institute of Quantum Optics in Garching, Germany. He tracks, using ultraviolet laser (UV) pulses, brief quantum leaps of electrons within atoms. These events are fast and last for about 100 attoseconds or 100 quintillionths of a second. How small is this time scale? The 100 attoseconds are to one second, is one year to 100 million years.

But even Kraus's work does not come close to the edge of time involved in such transitions. Attoseconds are too large of a time interval when one compares to the temporal realm known as the Planck scale. This scale marks the edge where distances and time intervals are so small that our very notion of time becomes questionable. Time became a scientific problem, nearly a century ago, when its absolute status was compromised by Einstein's special and general theories. According to the theory of relativity, the past, present, and future could only be relative. In fact, scientists are realizing that the best way to think about quantum reality is to give up the sense of time, implying that the fundamental description of the universe must be timeless. In fact, time, as we know it, is nonexistent.

It is quite surprising and contrary to the calibration of our rational mind. It thinks absolute time is a worldly reality. However, there exists no time in the empty space of the universe. Time's presence is only relative to the existence of matter. The clock runs in the spectator's mind that perceives matter and its related dimensions of fixed space and time.

So "now" is relative and depends on the observer. According to Einstein's special relativity, an observer moving at the speed of light will not experience time and will be timeless or immortal. This means that photons, traveling at the speed of light, never experience time, or a photon observer that started at big bang event and is arriving to us now, for him, there is no time-lapse. Today, we can prepare two quantum bits or qubits for short, in an entangled state in a manner that the information in the entangled quantum system (our two qubits) is distributed so that neither qubit on its own carries any information—all of the information is encoded jointly between the two qubits. Let us deliver one of the qubit to our classical friend Alice and the other to Bob. And the correlation of the two qubits is the secret they share between them. Unlike classical correlations, this highly unusual character of quantum correlation is "monogamous," in the sense that if two qubits are maximally entangled, then no other system in the universe can be correlated to the state of those two qubits. In other words, no one can eavesdrop on their secret and therefore can have no knowledge about what Alice and Bob are hiding.

Additionally if Alice and Bob wanted to see the secret resting in the quantum vault of the two entangled qubits, they could measure their qubit no matter where they are. Whatever result Alice gets will be the corresponding correlated result that Bob gets. In fact, weirdness of the quantum world is such that Alice and Bob could be literally anywhere in the universe. For example, Alice could be on the Earth and Bob on Mars, the two qubits will instantaneously communicate and convey the same results. As I described earlier, this is what frustrated rational mind of Einstein as he called it "spooky action from distance." Any sense of "now" is at best spooky as far as quantum physics goes.

It is only when, the observers are at moving at speeds that are slower than the speed of light, the observations result in the experience of time, a reality which is, once again hard to comprehend by our rational mind. It cannot tell whether this is a recipe for immortality or just another stubborn weirdness of the quantum universe.

Niels Bohr institute or Copenhagen interpretation acknowledges that there is a deeper reality. It transcends our everyday space-time reality of our world. The Copenhagen interpretation states that we create reality by observation, and there is no apparent reality without observation. Our senses are calibrated to give us an impression of a material world but that this reality is a reflection of something of a different nature. Does it mean that the universe is an event inside a conscious observer's mind?

It brings us back to our search for the nature of our true reality. Are there gaps between what we mentally grasp or mathematically describe about the world and what we sense, feel or experience? We know that no account of water, no matter how rich or mathematically true, can lessen our thirst. We also know, no matter how much one appreciates the abstract beauty of Maxwell's electromagnetic field equations or Einstein's famous equation $E=mc^2$, it turns pale by comparison to appeal of experiencing a sunrise.

Even the most sophisticated theories of the origins of the universe arising out of our collective comprehension seem somehow lacking that substance which experiential reality is made up of and in the end leaves us unsatisfied. In the words of physicist Stephen Hawking, "What is it that breathes fire into the equations and makes a universe for them to describe?" What is the nature of our experiences such as color, taste, sound, with which each of us is intimately familiar? How firing neurons in the brain can result in these subjective qualitative states we know as experience.

Clearly, an observer with mind plays a key role in experiencing and modeling this reality. Is the physical universe we see around us as mere flicker of digital information projected on the display of our consciousness represents our true reality? Or is there a deeper experiential reality?

Our new knowledge based on quantum computation and information processing implies there is. It is experiential and deeply rooted in mathematics and quantum nature of information. To access it, we would have to transcend our solid, physical, and material view of our universe and embrace a worldview based on computational, informational, mathematical, or abstract nature of being. A dance of bits, bytes, and qubits incredibly intricate and choreographed by none other than a grand abstract mathematical cosmic mind or a digital divine.

Hindus were first to describe god as a dancer, the Natraj. The universe, according to them, is an incredibly intricate dance, weaving in harmony and the mood of the Creator and its creation as one. All events interconnected influencing each other. Our modern view is turning out to be not very different. It is the cornerstone of our new reality. We are not separate. We are all connected at the deepest level through that one and play a small part in a truly grand dance we call our universe. All sense of eternal love, empathy, and compassion arises from this wholesome foundation. Whether, one constructs a reality based on the dancer or the dance, it is observer dependent. The nature and the rise of self are at the root of all forms of reality, abstract or physical. How does a separate conscious self arise from this grand oneness?

———— ∿⌁☯☯⌁∿ ————

Consciousness
Rise of Self

Conscious or Unconscious

Is consciousness a black and white property of life?

The answer is "No". Not only does a living being can exist in an unconscious state of being but it is capable of displaying many altered variations of the state of unconsciousness. Actress Natasha Richardson died following a minor head injury suffered while snow skiing. Tony Award-winning actor was only forty-five and lived in Manhattan, Millbrook, New York. According to reports, she fell while skiing and hit her head. She never lost consciousness. In fact, Ms. Richardson declined medical attention because she felt fine. However, an hour later, she began to experience severe headaches and then lost consciousness. After she was taken to the hospital, she experienced total brain inactivity or what has been described as being "brain dead."

Traumatic brain injury is caused by impacts that leads to widespread damage in one or more areas of the brain. A period of unconsciousness follows in many cases. The state of unconsciousness may last a few hours to weeks or even months. In most mild brain injuries, there may be only very brief alteration in consciousness, perhaps a feeling of being dazed. Ms. Richardson's case was one of the rare cases where a mild brain injury led to prolonged unconsciousness eventually leading to her death.

Coma is another state where victim is alive but loses consciousness. It is usually the consequence of acute brain injury. Comatose patients do not even experience sleep-and-wake cycle. They are always unconscious. For many patients, it is usually an in-between state. They may fully recover or may suffer a chronic neurological impairment leading to death.

The most familiar state of altered consciousness, however, is sleep. It is a state between being conscious and being unconscious. It is biologically essential. Our brain promotes it. This may be because some brain processes are best carried out in an unconscious state. One such function may be consolidating memory. Is our brain either conscious (on) or unconscious (off), or there is an in-between state? Rarely the transition from unconscious to conscious is reported to be like turning the on/off switch. More often, it is a gradual transition. It is initiated with reduction in unconsciousness with fuzzy or halfway states leading to being conscious. About half the participant of a study reported to being in this in between state, much like a quantum state of superposition, as they were intentionally awakened even from the lightest stage of sleep as determined by EEG monitoring.

Many of us have experienced the meaning of the phrase "put to sleep" with a general anesthesia. The anesthetized state of brain, in some ways, resembles sleep state of the brain. In this state we do not experience pain. We have no access to our memories. The brain waves, however, are present. Interestingly we lose track of time as well. If you were in surgery for hours, put to sleep by anesthesia, time did not exist for you.

Rene Descartes said, "I think, therefore, I am." Perhaps more appropriate is "I am conscious therefore I am." To be enlightened is to be awakened to your true, conscious state. The consciousness is intimately tied to the very existence of self.

Conscious Self Separate from Others

The universal informational entity, the whole stays whole unless an observer is born with a mind that is sufficiently different than the cosmic mind to be able to incorporate separation in its formulation. If there is no observer of this quantum oneness, it stays in state on being one, as quantum mechanical waves of potentialities in an ocean of silence. Only through the eyes of an observer, the waves of potentiality transpose into matter giving rise to another paradigm of reality marked by separateness and boundaries creating our physical universe. A reality that we observe, with our senses, interpret it with our local mind as we become conscious of our physical selves. I am different from you. This widely recognizable human perception has very deep roots. It is, in fact, one of the most fundamental natures of the cosmic mind; one is truly a form of many. Or one is hidden in many and many are hidden in one. It is the reason why a self even exists. Consciousness separates the self from the whole. One may understand this process to be like a mother giving birth to a child. Alternatively, a larger organization creating a smaller operating division. Or a small town is created from a larger county or city. The newly created entity will be a part of the larger original entity through certain relationships but will develop its own intelligence or mind to take care of certain local functions.

If self is conscious, what is it conscious of? It is conscious of its parts. It is also conscious of what are not its parts or rest. It is also conscious of the whole. The nature of self is the emergent property of its interactive parts that constitute the system. Myself is aware of its nature. It is aware of its needs and desires. It is aware of its boundaries. It sees colors; it sees matter with different properties such as hardness, flexibility, toughness, flow ability, transparency, and toughness. These myriad of properties create an appearance that the objects that surround itself are separate from each other, and the self is separate from everything else. From this point of view, myself is conscious of its separateness from the rest. The self experiences the world through its senses or sensors. All the senses conspire to provide the sense of boundaries, which define its physical reality.

When we look at the universe at the quantum level, the separation is not easy to define. There is no color, no tactile differences of hardness, flexibility, transparency, and no boundaries. There are enormous gaps between electrons racing around the nucleus inside an atom. The information entity, the whole comprises of quantum entities that are connected in myriad of ways locally and nonlocally pretty much like one entity, until an observer of the whole is recognized. The observer of the whole is undoubtedly part of the whole but has developed consciousness of its separateness from the whole.

Danah Zohar, in her book, The Quantum Self, describes the quantum level of our bodies and our consciousness. She writes,

> We conscious human beings are the natural bridge between the everyday world and the world of quantum physics . . . (There is) the possibility that consciousness, like matter, emerges from the world of quantum events, that the two, though wholly different from each other, have a common "mother" in quantum reality. If so, our thought patterns, and beyond that our relationship to ourselves, to others, and to the world at large might in some ways be explained by, and in other ways mirror, the same laws and behavior patterns that govern the world of electrons and photons.

Louisa Walker concludes that quantum field of possibilities is the spiritual essence of us. Physicist David Bohm suggests that mathematically the universe is similar to a hologram which deals with frequency domain and varying waveform properties. The holographic notion of consciousness has psychophysical implications. Bohm remarked that our views of reality are conditioned on lenses such as eyes, cameras, or microscopes, which focus, objectify, form boundaries, and particularize. Lensless holograms are distributed, lack boundaries, and are "holistic."

Experiences reported by mystics, especially in meditation or Samadhi, describe the loss of spatial and temporal boundaries and the emergence of the grand feeling of being one. It bears a remarkable similarity with holographic ("fractal") characteristic of the whole, which is also represented in every part. "Many know that there are many drops in the ocean, but very few know that there is the ocean in the drop, I want you to be those few ones," a quote from Prem Rawat to his followers, a leading spiritual master of our times. One may argue whether mystically induced perceptions are aberrations or clarified reality, but certain properties of holograms bear some likeness to consciousness.

In most spiritual practices, specially related to meditation and Samadhi, dissolution of separateness and the experience of oneness is the experience of divinity or enlightenment. This experience is so incredible, and yet only a few have experienced. So clearly, it is not easy to get to. I will discuss this, in details, in the last chapter, when I will talk about enlightenment. Words such as divine intoxication or divine juice are used to describe such experiences. In Hindi language, the word "tari" is used to describe one such state of mind. The English translation of the word would suggest the meaning to be a sleep state. Nevertheless, in fact, the true meaning is the consciouslessness or hypnotic or utterly intoxicated state of consciousness. Attention with qualities that a supreme lover showers on her loved one. The attention that is focused on one, just one, the supreme love, God. Everything disappears except the lover. When an elevated consciousness is able to witness the divinity (the whole) with such an attention and intensions, enlightenment results.

Hindus also describe another state where the consciousness gets elevated to the point where it crosses the interface of separateness and enters the divine entity in its completeness at the same time and still conscious of its separate self. From this perch, consciousness means pure awareness encompassing two essential qualities, knowing and clarity. The knowing is to know without reasoning or rational mind. Clarity is doubtless state where what is known is absolute truth. Such experiences are described by Buddha and Mahavir as they become one with divinity in the deep state of Samadhi. Now they could see and experience what the computational entity whole would know and experience. Such an extraordinary state of being, all knowing and everywhere, all times, is described by their followers. All concepts of spirituality, religion, and divinity are ultimately connected to such experiences of consciousness, in my view.

Consciousness in the sea of potentialities of cosmic mind represents the onset of the evolution of the self. Informationally, it split the whole in parts. The self with the formulation of separateness was on its way to evolve in the womb of oneness of the cosmic mind. It needed two additional computational structures to make a modern human. These were emotional and rational minds. The development and the evolution of these two minds further deepened its sense of separation. It could not, however, eliminate wholeness from its formulation. The self turned out to be a tale of three minds: cosmic, emotional, and rational, mediated by consciousness. Our sense of modern humanity is rooted in this nature of self.

———ᴡᴡᴏ◦ᴇʀᴏ◦ᴏ◦ᴇ◦ᴏ◦ᴡᴡ———

Chapter 3

Self—
A Tale of Three Minds

Nature of Myself
Humble Beginnings

Most of us do not remember the humble beginnings of our own self. Conceptually it is difficult to envision that we started our existential journey as a tiny cell hardly visible to the naked eye. It is also hard for us to imagine ourselves in our mother's womb for nine months waiting to be fully formed. Even harder to imagine is our journey initiating from our first home inside the tummy of our mother to the visible universe. We displayed incredible trust in strangers that we later understood as parents or guardians. We had no capabilities of doubting and we did not. In that humble and simple existence, truth was with us. We did not or could not sugarcoat our emotions. When hungry or thirsty, we cried until we were fed and felt full. There was no sense of social protocol, and we pretty much followed queue from our body. As naked as helpless and as truthful our self was, we had no idea that we will become what we have become today. Despite of all the pictures and movies capturing most of those occasions, we tend to forget or disregard that period of our humble existence.

Most of us do have some memories from our childhood. We recognize that learning was a crucial part of our lives as a child. Even though our parents tried to convince us that childhood was the best state of our existence, most could not wait to be an adult. We wanted to grow up to be someone. There were many suggestions. Be a doctor, lawyer, president, CEO, or a teacher. The list was endless. Mother always wanted me to marry a lovely young woman and have a terrific family. She made that clear repeatedly until I understood. Whether it is family, friends, relatives, organization, city, country, world, humanity or life or universe, we learned to see ourselves as contributing individuals and part of a myriad of larger organizations. We were conditioned to see an individual self or ourselves as one unit.

In 1927 Ludwig von Bertalanffy first proposed the human organism should be treated as an open system. An open system may interact with its environment in a growth or balancing fashion. Human body comprises of over 50 trillion living cells. Each of these cells is also open system aware of its environment through a network of molecular sensors. These cellular open systems perform collectively to cause the human organism to have a motion and various other functionalities. From a systems perspective, one may describe the self that we have conditioned ourselves to be as an emergent property of human system. As the emergent property, the nature of self is different from mere collection of 50 trillion cells. For example, self might enjoy a game of poker while 50 trillion cells may not appreciate sophisticated moves of poker. Consciousness is the basis of this Self.

Myself for Me or for Others, the Conflict

What is the nature of the self that is conscious?

The nature of myself is centered around two basic tendencies. First, I am for myself. Second, myself is for others. This duel nature of self is almost like the nature of quantum entities such as photons or electrons, which exist in dual state of being wave or particle. This duality leads to conflict in our mind. It also matches with the duel nature of the cosmic mind itself. One forms many, and many form one. The action of robber, in the headline below, pictures a deep conflict within, a conflict that relates to the dual nature of our self.

Calif. robber confesses to priest, turns in cash

Police still seeking identity of the thief who vanished after handing over $1,200

Tuesday, July 28, 2009

WALNUT CREEK, California—A man who robbed a San Francisco Bay Area bank apparently felt guilty enough to confess his sins and hand over $1,200, but not enough to turn himself in.

Police say a man went to a church Sunday night to confess that he robbed the Walnut Creek branch of the Patelco Credit Union last week. After expressing remorse, the man handed the priest $1,200 and left. The priest then called police.

Walnut Creek police Lt. Shelly James says authorities don't know the man's name or if the money he turned over was the entire amount lost in the robbery. The money will be returned to the bank.

Police say the man entered the credit union Thursday afternoon and gave a bank teller a note demanding money. He claimed he had a gun.

James says police are still looking for the robber and will arrest him when he's found.

What gives rise to this duel nature of self? It took me several decades to recognize the dual nature of my own self.

Seema and I have been married for twenty years. Just like most marriages, our marriage has been an instructive exercise in getting to know each other. I was born in a small city in India called Aligarh in 1960. Most of my childhood was spent in Delhi in a typical Indian middle-class family. My father, who retired as a professor of business management in a prestigious university in India was then a lecturer in a college in Delhi. With my mother not working, five children, my father had to work hard to just simply bring money to fulfill our basic needs. My mother came from a wealthy family. She could not see her children being deprived of basic wants. She always felt that she could work and support the family with extra income. However, times were such that my father felt the role of a wife was in the home, not outside working and bringing the bacon. In fact, women working outside were a social taboo in the Indian society of those times. My mother was one of the exceptional women of her times. She was educated and brilliant. She was a debate champion and distinction holder throughout college years, in times when most Indian women did not get any college education. My father was also exceptional in studies and topped throughout his college education. My parents married through an incredible alignment of fate. In a typical Indian arranged marriage style, both went through many potential suitors and eventually ended with each other through divine intervention. In India, all matches are made in heaven. There appears to be a deeper truth behind that saying. We grew up in a household where our mother and father constantly argued about finances, and my mother constantly wished to work to raise our living standards. My father held the line throughout, and my mother never worked. She died in 2005 leaving incredible impact on my eventual personality. I grew up with incredulous need of not be financially mediocre.

I am fourth of five children in my family. I have two older sisters and one older brother. Just like my parents, I turned out to be the one good at studies. For some reason, I was naturally inclined toward science and mathematics. I studied most of time and got admission in Indian Institute of Technology (IIT), Delhi, a prestigious institution for engineering education. I came to United States in 1982 for graduate work at University of Southern California (USC). Lure to penetrate the glitter of material wealth could not be greater anywhere but here. This land of opportunity is such an incredible engine for wealth creation that everything else such as skin color, accent takes a backseat. Money is like divine here. What creates wealth is almost always more important than what does not. I engaged myself in that pursuit like most others. I was in a rat race.

Rat race is an amazing race. Winning is more valuable than anything else. Winning promises material rewards. These could be significantly greater than bare financial security. For example, Bill gates, once the richest man on earth created wealth that would be enough to feed many generations of Gates to come. Rational mind is the hallmark of the rat race. Apparent determinism, certainty, and result orientation are the essential ingredients of this race. Every rat, in the end, is for himself. Power struggle, fear, and anxiety are common and prevailing emotions. I was in that mode. I was determined to do well, for myself. I sought my benefit in everything I came across. I judged everything with that standard. I was not trusting. My body showed signs of fatigue from severe negative emotions; unhappiness was rooted in my psyche. This new paradigm for me was different from what I grew up with. As a child, I recalled I was trusting, social, respectful, and simple; unlike what I had become. My child was still alive in me. It was fighting for what it thought was right. The kindness, love, empathy, friendship, freedom, selflessness, and truthfulness were uncompromising traits. These were not the hallmarks of winning a rat race. Not much made sense. My inner conflict was at paramount. I was in search of meaning of this existence, meaning of myself. Being a science person, I thought I will find answers in science, but decades of search in the scientific literature could not take me all the way to my true reality, my true self. This awakening came later from subjective experience of my inner self in the deep state of meditation.

Seema came to United States in 1989. Her concept of herself was rooted in her heart. She listened to messages that surfaced from her heart and gave them priority over mind. She always cared more for others than herself. I did not recognize her goodness right away. In fact, we struggled for consensus for many years. As I struggled to get answers that would transcend me beyond the powerful paradigm of the rat race, using science and spirituality mainly through many of the self-help books, it was becoming clear to me that myself had two distinct natures, both very real. One catered for my own self-interest, and the other was selfless or catered for others interest. What could lead to such opposite nature of self-residing in one body? My search led me to an answer, from a surprising source. The answer came from the fundamental nature of quantum computing. A new science of information processing that has a potential to combine all science, mysticism, and religion. It provided a clue to explain the dual nature of self.

Computationally Astute Informational Self

At the deepest informational level, living cell is composed of trillions of quantum variables, qubits. At the physical level, the quantum information stored in these quantum switches or qubits combine in the form of atoms, informational molecules, and other nano structures. Living cells form tissues, organs, and finally organized as life-forms including the human body. Therefore, the life forms such as humans are large macroscopic structures, which are buried deep in internal quantum features. From a top-down perspective and as a macroscopic body, humans would be closer to a classical computing or binary processing. At the root level, these structures are computationally astute to take advantage of both binary and quantum mode of information processing.

From a bottom-up perspective and as a quantum inner body, we would be closer to quantum processing. From this perspective, we form a seamless oneness with everything else in the universe. From the top-down view, derived from binary computation, we form a perspective of being separate from the rest. The consciousness is the interface of these two perspectives. On one side of consciousness, there is this grand oneness, whereas, on the other side, the separation with the rest.

Myself is fundamentally an informational entity whose fundamental aspect of being is the ability to know and act per guidance of a mind influenced by these two perspectives. Many programs drive my conscious self. These programs live in my mind that finally guides myself. My mind directs all the actions of bits and qubits through these programs to cause various experiences of self. Everything that myself does, therefore, can eventually be described by a series of interconnected tiny quantum of action or the qubit or programs governed by the interaction of one or more aspects of my minds.

"C'mon, Papa, we are not like computers with programs running inside of us," my eleven-year-old son Monish exclaimed as he heard of my newfound wisdom.

"Maybe not the kind that we know and are used to, but we compute incessantly and information is also at the root of our being." I was not sure if it consoled him.

Most of us do not see ourselves from this view. Our image of ourselves is mostly based of perceptual reality of our senses, consistent with laws of classical physics. We see ourselves as large solid three-dimensional physical beings where energy breaths life into our material body, separate from everyone else. Our top-down view based on binary computation makes us aware of this separation.

The separation leads to incompleteness, the product of which is desire. The most basic desire, common to all life, is of food, or a source of energy. Life is like a toy that operates on batteries. Food is the battery that all variety of life needs. Safety is another desire that is innate and results from preservation of the form. The separation need takes on a wide variety of forms to be complete. The self has evolved to have a free will to pursue the fulfillment of the needs and desires. Humans along with a few other life-forms have evolved a capability that enables a rational mode of processing information using our rational mind. It is physically contained in and informationally synchronized with the emotional mind.

My physical body is my emotional mind and is eventually rooted in the cosmic mind at the quantum level. The emotional mind provides the information network involving the molecules of information and emotion that facilitates communication from nonlocal regime to the self. In other words, the inputs from the cosmic mind are mediated to the living entity through the emotional mind. The emotional mind also links information circuits that allows the decisions from rational mind to reach all part of the physical body. So what I call as myself is interplay of these three minds. Which mind I use is a matter of what I feel is important and where I put my attention.

The nature of myself reflects a tale of three minds—emotional, rational, and cosmic minds—that work in concert, resulting in the experience of the self. Therefore, even though I am physically limited, I am unlimited from informational perspective as myself linked to the cosmic mind spans the whole universe. Each of the three minds contributes differently to the nature of myself by altering the computational mode of this informational entity. Since the nature of the three different minds is decidedly different, their influence results in different nature of the self. Two of these natures of the self are recognizable and present in all of us. First is centered on myself being separate from others, and the second one result from myself as one with others. If you recall from earlier discussions, these are exactly the two basic tendencies that make the cosmic mind or the whole. So from this point of view, self is made in the image of the whole.

Each living cell in my body is a self in its own right. It has made a commitment to support my being. My desires are imposed on my cells through a top-down network of informational commands. My cells obey to fulfill these wishes. Many times, my self-interests will put extra strain on my cells. My cell's functionality depends on it being healthy. Just like I feel tired or sick when my system is overused or abused, each of my cells feel the same through network of informational signals that flow throughout my emotional and rational brain. Each of my cells is also rooted in the cosmic mind and receives signals from that mind in the form of waves of guidance. If signals from rational mind and the cosmic mind do not match, my cells are often in conflict. Which proteins sensors and gears would need to move and in which direction? The conflict leads to internal pain or suffering. Since the programming of our modern, rational mind is often at odds with the functionality of the cosmic mind, this suffering has become a normal feature of our existence.

When rational mind is in a state of surrender, our decisions are made through the cosmic mind. This leads to ease of conflict within and thus a sense of peace or calm spreads throughout the body. The rational mind also can make decisions that can benefit the community or others. In this case, the rational mind works in concert with the cosmic mind as they coconspire to create the reality. Therefore, whether myself operates in a state of surrender to the cosmic mind or works in concert with it, the resultant processor is quantum in nature. I would call it our "quantum self." It is also our bottom-up perspective.

After its origin from quantum information processing, it appears that life has been evolving toward this binary mode of information processing. Our free will allows a choice between the queues from the grand cosmic mind and the rational mind. Through intentionality, we can choose one or the other. When we choose the rational mind over the cosmic mind, we act with our rational mind. This leads to second of the two essential natures of our self. The information processor dominating this nature is binary. I will call this as our binary self. There is a strong temptation for the rational mind to make decisions so that they benefit the self only. This mode of action leads to development of egoic entity, which is highly prevalent in our modern societies.

A Case of Mean Bacteria

Robert Austin of Princeton has been studying the social behavior of bacteria in order to help understand the social dynamics of other organisms, including humans. He has intriguing results about selfish and altruistic individuals and the social dynamics between the two.

Austin and his collaborators have found a single gene that controls bacteria "selfishness." If it is off, bacteria slow down their metabolism and reproduction rate when they sense their environment has been depleted of nutrients. This prevents them from entirely destroying their living space. However, if this gene is turned on the bacteria go right on eating until nothing is left. They even develop the ability to feed off other dead bacteria.

Interestingly, the gene is off by default when bacteria are found in the wild. However, if one puts these in a petri dish, and cut off their food supply, the selfish mutants emerge in matter of only days. These mutants consume all the remaining food, including each other, and then starve.

The experiment shows: first, the selfish bacteria win, and then everyone loses. This is a situation where selfish seem to have the advantage over the self-restraining altruists, but if everyone is selfish, then everyone is worse off. These results, however, could not be reproduced in the wild where selfish bacteria would not emerge. The bacteria in the wild exercise restraint, so there must be something different going on in the wild than in the petri dish. This is a fascinating example where a simple organism like bacteria is informationally able to change the nature of its "self" from quantum to binary depending on the influence of environment.

Modern human device contain both binary, egoic or selfish, and quantum or altruistic self in one body. The result is a conflict or internal unrest that is the prime cause of suffering in modern humans.

———ᴡᴡᴏᴏᴇ⌀⌀ᴇᴏᴏᴡᴡ———

On Being Human

Chapter 4

I, Mine, or My
Binary Universe
Humans
In Modern Times

My Binary Self—I, Me, and Mine

Obsession of Pressing Buttons

My two sons Vinay and Monish have been a tremendous source of learning for me. Among many intriguing things they did from their childhood, one clearly stands out; their obsession with pressing buttons. It may sound silly, but both I and my wife, Seema, was surprised when we first became parents and our first one Vinay was genuinely attached to this modern act of pressing the buttons or turning the switches on or off. The obsession continued with the second one Monish, who was equally obsessed if not more, with pressing buttons. We, in fact, named him "button Pressu" when we found his passion reaching levels that was apparent to guests who came to our house even for short periods of time. Now, that we had started taking notice of this act, we started to realize that almost all the little kids were obsessed with the same act. The toy designers evidently knew this nature of little kids well, and many popular toys were designed around that tendency. As my kids grew and moved on to video games and played some of them with incredible obsession or even addiction, what I realized was that it was the continuation of their childhood passion of pressing buttons. Evidently, video game designer were also acutely aware of this nature of our mind. What makes us so attached to this simple act? Why is a child's mind naturally obsessed with this act? The passion is not of pressing the button, as I found out, but it is what happens when you press the button. It is a fundamental principle of cause and effect, the hallmark of binary mind. Our binary mind calibrates itself according to the results obtained. It, in fact, reflects a fundamental binary nature of our self. Immediate response to an act such as either/or, yes/no, action/inaction, I am/or not forms the inner informational architecture of our binary mind. Video game designers tie this action to winning or losing and make this programming even more amusing.

Rational Decision Maker

Making decisions are the hallmark of our rational mind. Rationality and binary processing are at the root of this mind. It is an untiring decision maker and problem solver. This is what we appear do pretty much all the time. From simple decisions to complex ones, some we make consciously and others, without much conscious reflection. What should I wear today?

Should I eat strawberries or nuts? How should I choose investments? How should I manage my car problems? It does appear that our present life is full of decision making acts. Better CEO makes better decisions. More successful people make better set of decisions? There is a lot riding on making decisions and making them right. This has been the evolutionary discriminator for different species.

If we cannot find the physical entity called self or the conductor within us, we can feel the informational, computational entity self, the decision maker within. Our rational decision maker lives in our head. It is the rational brain. Every binary entity in our universe takes part in making yes/no decisions. It is the most basic form of rationality. All more complex rational thoughts arise out of this simple binary computational foundation. I must remind you that it only implies either yes or no, but not both at the same time. It can be restated as if we are separate than we are not one. In cosmic schemes of things, role of binary mind is such that it sees separateness, boundaries, and individual self or selves. Collective waves of yes/no decisions lead to simple to most complex information processing events leading to the creation of complex binary information structures and behaviors.

"We humans, in general, take pride in our capability of thinking of purely rational thoughts. Over time, our rationality has come to define us. It is, simply put, what makes us human. There's only one problem with this notion of human rationality: it is wrong," points out Jonah Lehrer in his latest book How Do We Make Decisions.

According to Lehrer, most of our decisions are a combination of the rational mind and emotional mind. The rational mind reaches all part of our body or emotional mind through an elaborate network of emotional circuitry. Action and reaction constitute a fundamental tenet of calibration of our binary mind. However, this calibration is not just limited to circuits of our rational mind. As the rational mind is being calibrated through the process of cause and effect, these emotional circuits are also calibrated. After the calibration is learned, these emotional circuits may trigger by themselves without any conscious queues from rational mind. Daniel Goleman calls these circuits as "low roads of decision making" in his book Social Intelligence. According to him, the rational circuits, "the high roads" of decision making, are slow decision makers. The speed at which the low road circuits function is extremely high, relatively speaking, he claims. This is a crucial realization as the rapid speed of emotional circuitry not only allows these circuits to respond to current rational thoughts but prepares itself for thoughts that are yet not thoughts. In addition, this circuitry also tries to keep up or queues pending list of tasks that are or have been on rational mind's portfolio of activities. Like a true high fidelity assistant, this circuitry calibrates itself, so it can help rational processor with every single one of its needs and desires. Our modern education system is primarily based on this reasoning. All of us practice this throughout our lives, calibrating our rational and emotional minds for various decision making acts that would help us to survive and thrive in our lives.

The binary thinking was not invented by the human mind. Human mind simply mastered it. The binary thinking is not necessarily detrimental. It is responsible for remarkable technological progress that has led to human safety and improved living conditions. In fact, if we had a mind of infinite capacity, capable of computing at infinite speed, there will be no difference between grand quantum processor, the cosmic mind, and the binary mind. However, we know that our binary processor is limited in power and speed, and therefore, it puts restrictions on the nature of the results obtained. Therefore, binary analysis often leads to aperception rather than the truth.

Our understanding of our world is through our senses. Through this elaborate calibration process, all of our senses collaborate to bring us the sense of boundaries which forms our perspective or view of our physical universe. We perceive movement, time, and space. We see colors; we see matter with different properties such as hardness, flexibility, toughness, flow-ability, clarity, and toughness. These myriad of properties create an impression that the objects that surround us are disconnected from each other, and we are separate from everything else. From this point of view, the existence of a separate self or an egoic entity is natural and easy to understand.

Rise of Egoic Entity, I Win—You May Lose
Separate from Others, Better than Others

Andre Agassi is one of the most brilliant men ever played tennis. He is also the most beloved athletes in history of this game. Agassi's incredibly rigorous training began when he was just a boy. By the time he turned pro, at sixteen, his genius had changed the game of tennis forever.

He struggled early on. He stumbled in three grand-slam finals. Then he shocked the world. He captured the 1992 Wimbledon. Overnight, he became a media favorite. In his recent autobiographical account, he recounts his glorious resurrection, a comeback with his impressive run at the 1999 French Open and his march to become the oldest person ever ranked number one. Being number one is the passion of the modern human existence. All sports we play are geared toward rewarding the winners. All corporations, societies, and institutions operate toward separating better from worse, a line of demarcation or separation. I am better than you are. I offer better services and products than you do. Or I deserve more than you. I deserve to live more than you. This extraordinary human obsession has extremely deep roots. It is, in fact, one of the most fundamental natures of the cosmic mind; one is truly a form of many. It is the reason, why a binary self even exists. Actions leading to survival and self-benefit are the hallmark of this binary self. It gives rise to an egoic mind.

Negotiation 101, What Is in It for Me?

A-One Home Shield (AHS) provides appliance insurance for home owners, business owners, and rental units. The company charges customers flat fees just to subscribe to their services. In case, an appliance is not functioning, the home owners call and schedule an appointment and pay fees for the technician to come and diagnose the problem. Seema and I decided to subscribe to AHS when we bought our new house thinking that it would allow us the peace of mind, one less detail to worry about. New house, new appliances, we did not need the service and paid for several years about $490/year. After ten years or so, some appliances started to exhibit signs of aging, and problems started to appear. Recently our washing machine just stopped working. We called AHS, which in turn sent a local technician, who is usually on contract with AHS. His decisions are generally biased due to potential conflict of interest. The technician arrived, in a great rush, noted the serial number, and assured us that it would be taken care of on Monday as the offices of AHS were closed on Saturday. On Monday, I followed up with AHS; they denied the claim, suggesting that the appliance broke because of abusive usage of the equipment. I was terribly upset to hear that as I knew that was not the case. Anyway, I came home talked to my wife, Seema, and assured myself that it was not the case. I then called AHS and let them know that the technician spent five minutes total time, and his opinion is biased and did not correctly diagnose the problem. The customer service representative heard me calmly and schedules another appointment for another technician to come to our house. This time the technician spent time looking into the unit; he discussed remarkably calmly with us that the unit has stopped working, and it will be taken care of. He assured me that he would call me personally and arrange an appointment to fix the appliance. In the mean time, Seema had to go out of the house to wash the clothes. This was causing her a lot of inconvenience. We waited, and for a week, no one responded. I finally got hold of AHS and asked for the status. They replied that the request was denied as the second technician's boss concluded that the appliance was misused. In addition, she informed me that they would not send another technician as they already had gathered two opinions. It was quite frustrating to me as I lost the temper and asked to be connected to a supervisor.

The above episode is a real-life example of a popular game taught to business students in various business management programs. It is called the ultimatum game. Here two students are asked to distribute $10 between them. The student that has $10 makes a ridiculously low offer to the other student. For example, he will offer $1. The other student expecting $5, if not more, suddenly has to figure out how to deal with such low ball offer. A common reaction is anger, frustration, and how do I get back at the other person. When played for just one round, results are heightened negative emotions. If the game is played for many rounds, the players are more likely to achieve satisfactory results. Without a doubt, real-life negotiations are more complex and involve rational and emotional mind in a wide variety of manners. Nonetheless, preoccupation of winning or getting better deals drives the negotiation process in a tug of war of two self-interests.

Our modern world is deeply rooted in the notion of negotiation leading to fulfillment of our self-interest. Most of us constantly negotiate from morning to night and not even know that. Most of the time these deliberations are steered to benefit self-interest. This leads to a special degenerative case of binary thinking where consideration for others is lost, and self-interest becomes of paramount significance. Myself becomes a self separate from others with a clear distinct identity. At the basic level, it represents a computational process in which the processor of the information (self) would like the outcome in favor of the self and not the rest or the whole.

The foundation of this processor lies in rational, machine-type Newtonian paradigm where more powerful machine will govern the less powerful machines. However, it is not just the rational processor; it also involves emotional circuitry that carries these messages to the rest of the body. In other words, the egoic decision making is not limited to just the high roads. In fact, the low road circuitry of the emotional mind are also calibrated in the pursuit to benefit the self. In the animal kingdom, the behavior that leads to living entities eating other live entities called food chain arises from this thinking. The byproduct of such thinking is a computational entity we call ego.

Ego: The ego in Webster's dictionary is described as the self as distinct from the external world. When we live from ego-only, we cut ourselves off from experiencing the world in a holistic way. Daniel C. Dennett, the philosopher perhaps best known in cognitive science for his concept of intentional systems, describes the ego as follows: "It is something abstract which amounts just to the existence of an organization which tends to distinguish, control and preserve portions of the world, . . . creates and maintains boundaries . . . "Me against the world"—this distinction between everything on the inside of a closed boundary and everything in the external world is the heart of all biological processes . . . a unified agent whose words they are, about whom they are: in short, to posit what I call a Center of Narrative Gravity."

The egoic entity dominates human behavior in our modern lives. It is the most recognizable self in today's modern society. We all have separate names, social security numbers, or other identifications. I am me. My name, my wife, my kids, my house, my bank account, my car, my, mine make all very clear sense to all of us. It is also a top-down thinking where self, e.g., me is reduced to a material entity without the possibility to existing in a super positional state. For large human self, it means using brain logic with no heart. At the extreme level, an egoic thinker would say yes to benefit for self and no to benefit for others or rest. The egoic thinker will agree for the benefit of the rest only if it benefits the self in some way.

The binary thinking is slow and operates at the speed of let us say, human decision making which may be in micro seconds for an exceptionally fast action. This is a sharp contrast to quantum processors that operate non-locally at almost infinite speed or instantaneously. The binary thinker operates in mostly serial manner and continually seeks knowledge to make better decisions that would lead to the outcome in its favor. It seeks other binary processors to augment its lack of complete knowledge. It seeks binary network to supply information. The binary processor is predictable and programmable. Once programmed well, an algorithm has no inherent uncertainty and the reproducibility of results is guaranteed. This is a sharp contrast from a quantum processor where uncertainty is inherent, and reproducibility of results is questionable, and all events are unique.

The binary networkers organize themselves to work together as one or more teams or groups: greater the number, greater the potential for power. A pecking order is critical as information and communication must be managed. A leader is often chosen that could lead the group, hopefully for the group benefit. Often group members receive dissimilar benefits. For example in a corporation, each employee is paid according to the perceived benefit it provides to the organization. Usually, higher pecking order comes with greater benefits, which keeps members aiming for the positions of higher authority. Once the egoic thinking sets in, a continual struggle between individuals within or beyond the group goes on at every level since binary thinker optimizes benefits in its favor at all levels. The fear is its driving force. Egoic entity is constantly thinking, watching, and judging. It is in this way that ego gives the self an illusion of "safety" within the thought patterns of a binary thinker. This is the root cause of suffering in human society today.

As described, in the beginning of the Bible, "Adam's eating the fruit of knowledge" led to the onset of suffering in human society. It, in fact, started Adam on the path of binary thinking in my humble opinion. Initially, it was a competitive advantage for humans and a tremendous step forward in the evolution of humanity. Humans were not physically stronger than many animals. However, directly through binary thinking, which became their most powerful weapon, humans were able to outwit other animals.

Then binary thinking grew further. Through successes realized from such manner of thinking, it became acceptable and even desired in human society. Gradually, more of our identity went into the mode of binary thought. Humans identified themselves more and more as rational thinkers. This continually isolated us from the depths of our quantum being. Our identity has been driven by egoic entities powered by self-benefit. I, my, me, and mine became more meaningful. I am better than others scoped more and more as the binary thinking took greater foot hold in human society.

After thousands of years of embracing the egoic mind, our greatest weapon has become our greatest net. Today, we observe the egoic thinking at its prime with all its adverse effects. More humans are internally suffering stress related ailments and living much compromised life as fear takes control of their lives.

Many governments and nations operate with a power structure heavily influenced by egoic thinking. Terrorism is an example of that extreme egoic thought process. People kill themselves, just to kill others. Such thinking is particularly amplified by the science and technology with development of advanced weapons technology. We would destroy ourselves and the earth, simply because the binary network has enough power to do so. It simply needs a reason.

Egoic Pleasures, Deals with the Devil

What drives egoic mind? Egoic mind centers on self-pleasure or instant gratification. So much so, it often engages in risky behaviors. It is not that egoic mind is not afraid of the consequences; in fact, he is usually extremely afraid. It is just that getting these pleasures or fulfilling his self-interest in the moment is so paramount that he loses sight of the risks involved. Others could be in pain, or he may have inflicted pain to others to get his pleasures or he himself may be adversely affected in the end as long as he receives these pleasures, in the moment, he may be okay with it. All these pleasures have the capacity to turn into sadness, pain, or other negative feelings in the long run. In other words, the egoic mind is naturally attracted to deals with the devil.

What is pleasure? The answer lies in a set of molecules of emotion; most significant of these is called dopamine. Yes, this is connected with the nickname for cocaine, dope. How does this work? This neuroreceptor dopamine plays a vital role in our sense of feeling good. It, therefore, influences our just about every decision. The human brain is continuously computing by igniting or electrically firing trillions of neurons. However, what is the purpose of all this work? For an egoic mind, it all runs around an addiction—the addiction to dopamine or feeling good. We are indeed wired to feel good. It forms our deepest and most fundamental informational circuit. The egoic mind is no different with exception that it seeks to feel good through acts that benefit his self.

Through evolution, animals have learned to be rewarded with dopamine by actions that promote self-survival. For example, eating, procreating, winning, and helping others all result in dopamine release. We know, when we eat food, we feel good. But it is not the nutrients that we receive that makes us feels good, it is the dopamine released as a result of doing these acts. Therefore, something within us keeps track of what we are doing and rewards each move. The egoic mind latches on to this entity and identifies all the activities that are eligible for instant gratification. Under threatening situations, egoic entities do quite well. Intent on self-preservation, through determinism, reductionism, and machinelike perfection and domination, egoic entities made themselves winner of the survival game. Brain rewarded this behavior with a generous dose of dopamine, and therefore the winning race, our human race, evolved with incredible focus on producing egoic entities. Over centuries, egoic mind has learned to give significance to behaviors or acts that lead to generous dopamine release.

This dependence of egoic mind on dopamine has led to a wide range of human behaviors, in our modern society. Simple behaviors like getting better deals to accumulation of material rewards such as wealth is all driven by such needs. Simple acts like collecting airline points, clothes, and shoes are considered normal, and most see no problems with such egoic behaviors. The problems arise when such pursuits are taken to extremes. For example, extreme egoic behaviors tend to be overindulgent in sex, food, gambling, video games, and various addictions like drug, smoke, and alcohol. Sexual acts, pornographic materials, alcohol, dominating others into submission—all lead to the release of dopamine, and the egoic mind tends to latch on to these acts. Many of these indulgences have become habits or way of life in our society and are simply passed on from generation to generation. A community with only egoic entities would have bleak future.

Egoic Entity—Power and Superiority

Egoic mind is naturally attracted to influence, power, and authority. Egoic mind seeks power and authority over others. It does it through reductionism, one of the legacies of Newtonian physics. Egoic mind seeks separation from others. It sees people as disconnected from one another. It sees differences in people rather than what they have in common. It doubts, judges, assumes, and classifies people according to their wealth, character, and appearance. It is the egoic mind, which is called evil in Christianity as well as in most other religions across the world.

Egoic mind thinks that it is better than others, and therefore its actions are directed for it to rise above the masses. It has incredible need for achievement. However, when one achieves something driven by an ego, one may feel good for a while, but soon it is replaced by the feeling of emptiness and sadness inside. It forces egoic mind to constantly seek more, just to feel temporary gratification.

Egoic mind not only feeds itself on superior achievements but also on negative emotions such as fear, lack of trust, and lack of self-confidence. Chronic negativity grips egoic mind in subtle ways, and one may not even be conscious of it. One's personality changes dramatically. It is as if one is controlled by some demonic authority or entity. And one actually is, in some ways, by this egoic entity. If you find yourself complaining all the time, consider yourself unlucky, victim, or the one that has anger inside pretty much all the time; you may be possessed by this evil entity. If you are, don't take offense to that, most of us are; and it is our modern upbringing and daily struggle that takes full responsibility of instigating this entity within our personalities.

Egoic mind views the world as a combat stage. It sees most situations as a race it must win. It spends mental energies thinking about the battles it wants to wage and stances it takes against the demands of the world. As soon as it senses that someone is asking something from him or her, it starts planning how it will arrange best solution suitable for its self-need. Egoic entities follow the egoic network and its hierarchy, although it always fights to achieve higher pecking order in the network.

Extreme egoic tendencies lead to narcissism. Throughout history, into modern times, many including dictators, cult leaders, and tyrants have assumed that it was their right to dominate others into bondage. The philosopher Bertrand Russell was perhaps only slightly exaggerating when he claimed that all men want to be God and that some cannot believe they are not. A narcissist may not be too far apart from a tyrant in his self-images of superiority. Narcissists think they are superior to anyone else. Narcissists believe that they are fantastic, exceptional, and expect others to see and celebrate their superiority. It is immaterial if they actually are or not or if there is any evidence to support it or not.

"So there must be some instance when you were wrong?" asked a colleague.

"I was wrong once," said Bill.

"Really!" surprised colleague that Bill at least admitted to being wrong once.

"When was that?" asked the curious colleague.

"It was when I thought I was wrong, but I really was not." replied Bill.

Egoic mind has effects on behaviors in subtle ways and sometimes in not-so-subtle ways. Egoic entities will generally not ask for help, advice, or directions as it would be beneath them and would mean admitting that they are less than perfect. Narcissists often think they know or should know everything, and if they do not, they will often pretend to. According to them, anyone who needs help is deficient and so is anyone who cannot survive on his or her own. Narcissists often do not admit making any mistakes, and therefore, they do not learn from mistakes. They would always find someone else responsible for their own mistakes. Usually, they do not learn from anyone since it is beyond their grasp that how can a lesser life have anything to teach them, a superior being.

Narcissists do not see that others may have the same needs and rights as they do. Others simply exist as objects of their desire or acquisition. It often leads to emotional personality disorders characterized by simple lack of warmth and kindness to complete emotional detachment as found in personalities of many serial killers or psychopaths.

The British serial killer Dennis Nilsen was born in Fraserburgh, Scotland, on November 23, 1945. He was the only son of Betty and Olav Nilsen. Betty, Dennis, and his two siblings lived in the home of Betty's parents.

He grew up in an unhappy household, full of conflict; his father's marriage ended after seven years.

As a child, Dennis was quite common. He never displayed anger, cruelty to animals or to other children. Nor did he exhibit any aggressiveness typically associated with childhood stages of many serial killers. He grew up a solitary, mostly neglected by his mother. Dennis could not hold any relationship for any significant length. He had one brief relationship that eventually fell apart. After one year, the killings began. Dennis was thirty-three years old. He met a young man in a pub. He invited him to his home. Both drank and slept together. The next day Nilsen realized that he was going to lose his friend as he was getting ready to leave him. He decided to kill him. He first strangled and then drowned him. Later reports showed that he felt no remorse over what he had done.

Several years later, Dennis was caught flushing dead body parts down a sewer. But until that time, his killing spree continued. Dennis was reported to be more concerned about the problem of disposal of the bodies than any of the feelings he might have hurt. In prison, he showed no empathy and had no conception of the enormity of his crimes against the others.

Like narcissists, psychopaths lack empathic attitude toward others. They see others as objects to be manipulated. Both psychopaths and narcissists demonstrate substantial awareness to grasp ideas, choices, needs, courses of action, and priorities. Both like instant gratification. Both want everything now. Their whims and urges take precedence over the needs, preferences, and emotions of others including their loved ones.

Psychopaths often feel no remorse when they hurt others. They often do not display even the most rudimentary level of conscience. They love to justify or rationalize their behavior through reason or intellectualization. The psychopaths firmly believe the world is a vicious and unforgiving place. The survival is the ultimate reward of this game and the only law is survival of the fittest. People are either "all right" or "all evil." The psychopaths extend their own vulnerabilities, weaknesses, and flaws to others and force them to behave the way they want them to. Like narcissists, psychopaths are full of anger, resentment, abusive, exploitative, and incapable of love or affection.

Both narcissistics and psychopaths generally make poor member of human modern society. They are ill-suited to engage in the give-and-take of any civil society. Many become criminals engaging in professional crimes such as identity theft, fraud, and con artistry.

Both narcissists and psychopaths do not honor obligations, are unstable, unpredictable, vindictive and carry grudges. As a result, such individuals can rarely take or hold a job, repay debts, or remain in lasting relationships. They are often passionate and believe themselves to be immune to the outcomes of their own actions. It is quite a challenge or almost impossible to reform them as their intense passion makes them immune to weighing decisions related to others needs. They are always in conflict with authority and often on the run. When such people are appointed in the leadership positions, sufferings result.

The combination of such personality traits and power is so incredible that often God had to take birth on this earth to eradicate such evil. At least, such is the assertion in many ancient stories. One famous Hindu tale is about a king called Rawana, an ancient king of Lanka with ambitious and narcissistic personality. Hindu holy book Ramayana is the story of how Shri Ram, incarnation of Lord Vishnu, battled him and restored the peace and harmony. Throughout the world, history is full of such dictators. One would think after so much of destruction, pain, and suffering at the hand of such narcissistic leaders, our modern society has figured out a way to keep power away from such folks. It is not true. In fact, it is suspected that a high percentage of corporate executives and community leaders are narcissists who have learned to hide their inner devil and project an image of a good leader. Whether that is actually true or not, I leave it to you. But it does make a good Hollywood story!

Egoic Mind and Collective Good

Game Theory, Bounded Rationality, Satisficing Egoic Entity

Self-interest and self-preservation is supreme to an egoic thinker. Egoic mind realizes the contribution of others to maintain its self-interest. Therefore, by sheer logic and rationality it can arrive at decisions leading to cooperation and collective good. Framework of game theory has been extensively used to analyze various economic and social situations involving egoic minds. Game theory is a collection of rigorous models where decision makers interact with one another. A popular game is the Prisoner's Dilemma. In this game, two players, partners in a crime, have been captured by the authorities. Each is placed in a separate prison cell. Each is offered the opportunity to confess to the crime. The game theorists often represent this game by the following matrix of payoffs:

	not confess	confess
not confess	5,5	-4,10
Confess	10,-4	1,1

The above matrix indicates four scenarios of possible payoffs. These payoffs, for each of the prisoners, are typically represented by a set of numbers representing individual share or utility. Please note that higher the numbers are, more utility or higher payoff the option carries. Game theorists like these numbers as they are able to put these in their computer models and achieve concrete numeric results. However, in real situations they could represent real gains. For example in our situation, the two prisoners have stolen 10 million dollars.

If neither suspect confesses, both will go free. They would split the proceeds of their crime. Let us denote this by five units of utility or five million dollars for each suspect. If one prisoner confesses and the other does not, then the prisoner who confesses would testify against the other. In exchange, he goes free and gets ten units of utility. Moreover, the prisoner who did not confess goes to jail, which results in the negative utility, let us say that value is -4. Now, what is negative payoff? It simply means that this prisoner not only got $0 worth of proceeds from their steal but also got prison time. So negative number is game theorists way of reflecting or modeling the pain, which is greater than just not getting the stolen amount.

If both prisoners confess, then both are given a lessened term, but both are convicted, which let us represent by each one unit of utility. It is better than having the other prisoner confesses but not as good as going free. Game theorists can alter these utilities to model wide varieties of situations.

Game theorists have been fascinated by this game for a variety of reasons. First, it represents many real-life situations. For example, we can calculate the results using "contribute to the common good" or "behave self-ishly" decision. It represents a common situation for us modern humans where we are constantly choosing to do acts that are in our self-interest or for community interest.

Second, this game can shed light on how an intelligent self-centered person would behave. A lot of modern humans fall in that category, so such models may predict their behaviors. Getting back to our game scenario, irrespective of his partner's decision, it is best to confess. If the partner in the other cell is not confessing, it is possible to get ten instead of five. If the other partner confesses, it is possible to get one instead of minus four. Yet the pursuit of individuality results in each player getting only one unit of utility, much less than the five units each that they would get if neither confessed. It is this conflict, whether to pursue the personal goals or the common good, which is at the heart of many problems that game theorists try to solve.

Third, the game changes obviously if it is repeated in the future. Let us assume this game is over. The suspects either are freed or are released from jail. They commit another crime and the game repeats. Repetition opens the possibility of being rewarded or punished in the future for current or past behavior. Game theorists models confirm if the game were repeated often enough, the suspects would be inclined to cooperate.

If one looks at the modern humans, it makes sense that we often run into decisions where we must choose between acts for self-good and communal good. Does it happen to other life-forms: dogs, cats, monkeys, bacteria? Many studies now show that, it is not only modern humans that are riddled with this conflict; it may be quite widespread in fact throughout the nature. Even single cellular organism can be riddled with the same conflict. The decision making behavior described by game theory also seems to apply to individual decisions made by bacteria in theirs colonies. Recent studies show remarkable success of these models especially under the circumstances where each bacteria needs to choose between self-welfare and community health. It is giving us indications of how similarities may occur at the informational level between simple organisms such as bacteria to highly complex organisms like ourselves.

Eshel Ben-Jacob, a physics professor at Tel Aviv University and a fellow of the Center for Theoretical Biological Physics coauthored a study published in December 2009 online edition of the journal Proceedings of the National Academy of Sciences. He coauthored this paper with three other scientists at the center: José Onuchic, a professor of physics at UCSD (University of California, San Diego, and a codirector of the center; Peter Wolynes, a professor of physics and chemistry at UCSD; and Daniel Schultz, a postdoctoral researcher at UCSD. The paper used the game theory framework of prisoner's dilemma to explain the complex interplay of genes and proteins that colonies of bacteria rely on to launch different survival tactics during times of environmental stress.

When a colony of bacteria is put under severe stress, such as famine, each bacterium faces a dilemma. First, it can create spores. In this case, the mother cell dies. But before its death, it stores a copy of its DNA in a spore. The mother cell then breaks loose releasing its DNA encapsulated in spore and remaining proteins to the environment. The spores remain in dormant states until the environment improves. At that time, the bacteria can germinate from spore state into a fully functioning bacteria.

The second way a bacterium responds to extreme stress is to get into a state called "competence." In this state, the bacteria change their membranes to allow the easy absorption of material from the dying cells and hope to survive even under these unfriendly conditions. When normal conditions are restored, bacteria return to normal life without having to make a spore.

In the colony, each bacterium attempts to resolve this conflict, whether to produce spores or go in competence. The advantage of competence is the ability of the colony to promptly returning to normalcy. However, if the conditions become even worse, the cell will die without propagation of its genetic information.

How does each bacterium make the decision? Each bacterium in the colony communicates with the rest by chemical messages and performs a complex decision making process using a specialized network of information molecules. Observations have shown repeatedly that about 10 percent of the bacteria enter competence. But how is it that such few make this decision and which cells take this chance has been a mystery.

"It pays for the individual cell to take the risk and escape into competence only if it notices that most of the cells decide to sporulate," explained Onuchic. "But if this is the case, it should not take this chance because most of the other cells might reach the same conclusion and escape from sporulation."

Prisoner's dilemma for bacteria reflects humanlike decision making dilemmas with one difference. According to Onuchic, bacteria usually do not cheat their friends or other colony members. They inform them by sending chemical messages about their true intensions. "Each bacterium must decide whether to become a spore; that is, to cooperate, or escape into competence, or take advantage of others, while it has a limited time to decide while a clock is ticking. We discovered that each cell has an internal timer whose pace changes according to the stress it experiences—the pace goes up for higher stress decisions such as in humans. Our internal clock speeds up under danger because of the secretion of adrenaline and, therefore, we have the sensation of time slowing down. Besides internal stress, each bacterium adjusts the pace of its timer accordingly to the stress of its peers and their intention to sporulate or to go into competence," said Ben Jacob.

Most classical game theoretical analyses predict that rational, self-interested players will make decisions to achieve outcomes known as Nash equilibrium. Nash equilibrium does not mean the best cumulative payoff for all the players involved. It is, however, an outcome from which no player can increase his or her own payoff one-sidedly. Often players improve their payoffs by agreeing on to strategies that are more cooperative. Often, decision makers are less selfish and less strategic than these models predict and respect social factors such as reciprocity and fairness.

Herbert Simon has been credited for sweeping changes in microeconomics and was awarded the Nobel Prize in 1978. Simon was influenced by the fact that entrepreneurs rarely followed the marginalist principles of profit maximization or cost minimization in running organizations. The profits were not maximized mainly because the complete information was not available. Simon believed that uncertainty about the future and costs in gaining information in the present limited the extent to which a perfectly rational decision could be made. Thus only "bounded rationality" stays as a viable choice, and one must make decisions by "satisficing," or choosing that which might not be optimal but which will make them happy enough.

Rational behavior in economics simply means that an individual behaves to maximize his utility function, of course, within the constraints they face. The term-bounded rationality means that the rational choice is limited by both knowledge and cognitive capacity. It is a key concept used in modern behavioral economics as local variables often force a cooperative existential outcome of a situation through bounded rationality. It provides a way to learn rational actions aimed at common goods without invoking truly altruistic behavior.

Human babies are born with a mind wired to love. All our modern education system is the training in logic. It is difficult to initiate child to learn logic as many of us parents know. Children instinctively do whatever it takes to move away from logic-based learning. John Nash courted unbounded rationality to the limit of the device, his own rational mind that allowed him the rationality in the first place. He paid dearly for it as his rational mind suffered severe crash. John Nash was awarded the 1994 Nobel Prize in Economics for his work analyzing the nature of noncooperative games. In A Beautiful Mind, a popular film, the character of John F. Nash played by Russell Crowe concludes the followings as he accepts his Nobel Prize.

I've always believed in numbers and the equations and logics that lead to reason.

But after a lifetime of such pursuits, I ask,

"What truly is logic?"

"Who decides reason?"

My quest has taken me through the physical, the metaphysical, the delusional—and back.

And I have made the most important discovery of my career, the most important discovery of my life: It is only in the mysterious equations of love that any logic or reasons can be found.

Binary Thinking and Empathy

Binary thinking arises from quantum thinking? It is truly a degenerated form of quantum thinking where the connection with others is broken or ignored. For humans, this connection is manifested in the form of empathy or compassion. All humans are born with it. The way of living and distinctive emphasis on egoic entity makes people lose or ignores these emotions as they progress through their lives. Science is making headways in understanding how these two types of thinking are connected to information networks involving specific molecules of emotions. Researchers at the University of California, Berkeley, have found evidence that people who are more empathetic possess a certain variation of the oxytocin receptor gene, also known as allele. All humans inherit a variation of this gene from their parents. This study found that the most empathetic individuals who were able to get an accurate read on others' emotions had two copies of the G allele.

Informally, known as the "cuddle" or "love" hormone, oxytocin, an informational molecule is secreted into the bloodstream and travels to rest of the body or our emotional brain, where it promotes social interaction, bonding, and romantic love, empathy among other functions. What happens when connection facilitated by such an emotional circuitry is lost?

We show no empathy. Without empathy, we lose connection to others. People who lack empathy are incapable of loving others as well. Generally we get to know others by listening, asking questions, observing, and being sensitive to verbal and nonverbal cues from them. Someone without empathy will not interest in anyone else's self. Without empathy, it would become easier to have a dismissive reaction to other's feelings, wishes, and desires. Someone without empathy can be painful to others as their feelings mean little or nothing. Such persons may see their partners as a possession or become unwilling to provide support, assistance, or tenderness when needed. They may not listen or simply not respect their partner's differences. Without empathy, we may lose usefulness in helping others or living our lives to serve others. It is getting clear that loss of this emotional connection to others is the root of a large number of personality disorders like narcissism or psychopath attributed to extreme egoic behavior as described before.

Like anyone else, narcissists or psychopaths may also have a genuine desire to be loved or cared by others. This desire remains often unfulfilled and hidden. It is because of their personality traits, which allow almost no one to get close to them. Psychopaths and narcissists are aware of the effects and can be genuinely saddened by their inability to control it. Their lives usually stay devoid of a warm, strong bonds, or stable social network.

According to Russell Ackoff, a systems theorist and professor of organizational change, the informational content of the human mind can be classified into the following five categories:

1. Data: symbols
2. Information: data that are processed to be useful; provides answers to "who," "what," "where," and "when" questions
3. Knowledge: application of data and information; answers "how" questions
4. Understanding: appreciation of "why"
5. Wisdom: evaluated understanding

All data without reason or meaning is random. Reason or meaning can only be assigned by a mind. It could be the cosmic mind or an emotional mind or a rational mind. Ackoff indicates that the first four categories relate to the past; they deal with what has been or what is known. Only the fifth category, wisdom, deals with the future because it incorporates vision and design. With wisdom, people can shape the future rather than just grasp the present and past. However, achieving wisdom is not easy; people must move successively through the other categories. One facet of Ackoff's hierarchy is temporal. He says that while information "ages rapidly," knowledge "has a longer life span" and only understanding "has an aura of permanence." It is wisdom that he considers "permanent." Ultimate wisdom of reasoning leads to love. Ultimate wisdom of doubt leads to trust. Ultimate wisdom of "I" leads to "we." Zeleny proposed to add "enlightenment" on top of the Ackoff's hierarchy. Enlightenment, according to Zeleny is not only answering or understanding why (wisdom) but attaining the sense of truth, the sense of right and wrong, and having it socially accepted, respected, and sanctioned.

Many of us relate to our binary rational mind as our true self. However, it will only amount to limiting the scope of our true self. The rational mind is a recipient of information from the cosmic mind through emotional mind. The rational mind has a free will to accept this information or command the emotional and, informational network to what it concludes as the superior decision. The binary self is complete only when it is in a complete state of surrender to the cosmic mind. This is a difficult wisdom to grasp, and every binary self progresses on its own journey to make this realization.

Emperor and Servant, Quantum Self (Heart) vs. Egoic Mind

I would like to share with you a story that Osho, Bhagwan Rajneesh, a spiritual master, often recited in his discourses. The story is about an emperor who had a favorite servant. The servant served the emperor for many years and finally, the emperor was very delighted with his services and asked him if he wanted anything.

"My lord, you have given me everything; I eat your food, I wear your clothes, I live in a great palace, and I have a great privilege to serve you. I honestly do not want anything."

"Nevertheless, we are very happy with your services. There must be something that you want," said the emperor.

"My caretaker, my lord, If you really insist, there is one thing that I like."

"What is it? Do not be afraid, say it," said the emperor.

"Well, my lord, please spare my life. I just wanted to see how it feels like to sit on the throne of yours and how does it feel like to be you, my lord."

"Oh, that is easy. Tomorrow for one full day you will be the emperor."

Therefore, the emperor sends the news to the whole kingdom that the servant will be the king for one day, and he will have all the powers of the emperor.

Just as planned, in a great ceremony, the emperor celebrated the service of his servant and made him the king for that day.

The first order that servant executed was to enslave the emperor and declared himself as the emperor forever.

This story is highly relevant for us modern humans. Emperor is our heart or quantum self, which used to rule our lives, and servant is our egoic mind, a part of our binary self, which has gained control over our existence today. It is at the root of all human misery.

My True Self
Us, Ours, and We
Quantum Universe

True Nature of Myself, Quantum Self

What is the true nature of myself? I would like to suggest that our true self is made in the image of quantum bit or qubit. What does that mean? It means that its true nature mirrors the nature and properties of qubit. How can a large human represent qubit? Of course, physically it is not possible, but from an informational perspective, it could be possible. Take an example of key and lock device. It is a physical large device but represents a simple binary BIT of (0, 1) information. Similarly, a large living entity like ourselves may be built in the image of a qubit and operate using the quantum information logic embedded in its true nature.

Bohm suggests that each part of the universe contains information of the whole universe. Since qubit represents the smallest and most fundamental unit of our universe, each qubit contains all the information about the universe. He believes the universe resembles a hologram. What is a hologram? Most of us have seen these on credit cards or three-dimensional images eerily suspended in space theme parks such as Disneyland. One remarkable attribute of holograms is that any part can create the entire hologram. All one has to do is to shine it with coherent laser light. In other words, each part contains information about the whole. This principle, says Bohm, forms the basis of our universe as well.

For Bohm, the world is an indivisible whole. The oneness or unity binds all in a way which is not easy to grasp by our perceptual reality. Just like the hologram, even the infinitesimal point of the universe contains all information needed to represent the whole. The entire informational universe is enfolded within each of its parts. "Many know that there are many drops in the ocean, but very few know that there is the ocean in the drop, I want you to be those few ones," a quote from Prem Rawat to his followers, a leading spiritual master of our times. It is the essential nature of qubit, which prevails in the universe, qubit is the drop and qubit is also the ocean. It is a part of the hologram as well as it is the hologram itself.

The essential nature of my true self follows from the following four fundamental nature of qubit.

1. Just like a qubit my quantum self always seeks the company of whole or divine

2. Just like a qubit my quantum self always like to be in state of oneness or selfless state of being.
3. Just like a qubit my quantum self is a doer. A binary state of being or quanta of action is a fundamental part of qubit. All its actions are surrendered to whole or divine.
4. Just like a qubit my quantum self likes to form one mind with others like cosmic mind.

1. My Quantum Self Stays in the Company of Divine

The Communication

The cosmic mind is engaged in grand quantum computing at every level from the tiniest to the largest or the whole. The whole stays in the most efficient computation state, which is the state of superposition or parallel quantum computation involving signals that take place at the speeds close to infinity. The self is embedded in the whole. It provides inputs or constantly transmits quantum information to the whole in local and nonlocal manner. It may provide high fidelity unprocessed information or may process some or most information locally and transmit the rest to whole. The self observes the whole from two points of view: one, from the outside, is the information about our external environment. This is like us learning about the world around us through our five senses. The other is observing the whole from the inside. This is inner experience through feelings using emotional mind comprising of informational molecules, which are present in all life-forms. This emotional mind is always in sync with the universal mind of the whole. It obeys the universal mind in the absence of directives from a rational mind, or the binary self.

The whole computes based on the transmission it receives from the self and communicates the results of this computation back to the self. In living beings including humans, such information is received in the form of feelings. This information that I have called as wave of guidance is constantly provided to self whether any acceptance or even acknowledgement is made. The self is, therefore, a constant recipient of the quantum information from the whole. It is up to the self to make sense out of the received quantum information. The message of this communication is universal. Being in tune with this wave is the central focus behind all form of meditative activities. In the depth of prayers, one eventually gets tuned to this wave of guidance. All religious leaders, saints, spiritual masters, and enlightened beings throughout human history have agreed to this communication and have named it as central to divine experience.

I met Teresa recently on a business trip to Oklahoma City. She has been living with Al and Ellen Fletcher. I have known Fletchers professionally and at a personal level for about twelve years. Fletchers are devout Christians, both with true saintlike nature. Teresa, now mother of two, had a turbulent childhood and was married at an early age and got pregnant immediately. Just like her childhood, her marriage turned out to be very tumultous. She essentially spent twelve years in abusive relationships with her husband. She was poor and had to take care of herself and her two children. She changed home twenty nine times from thirteen to thirty seven years old, before she had met Fletchers. Fletchers gave her home to live. By her own account, her life turned around after JESUS saved her about twenty years back. She ended her relationship with the abusive husband and today has managed to educate herself. At the age approaching fifty, she is peaceful, happy, and looks forward to living every single day to fullest. I asked her the essential difference between before and after she was saved. I was very surprised by the clarity of her response. "Before I was saved I was the only one living in my body, after I was saved, I know besides me God lives inside me. Now I involve him in all I do and all that happens to me," she responded with incredible simplicity and truth. "So how do you communicate with the God living inside you?" I was very curious. "I simply talk to Him and simply listen. Sometimes the knowing just comes to me or other times I get signs from Him," she further explained.

"This is quite a common experience, once you are saved by Jesus," explained Ellen while trying to search for an appropriate verse from a small Bible she always carries with her. Ellen, from her own admission, was saved when she was just twelve years of age. "When you have that deep relationship with Jesus, you not only hear, but you also see visions," explained Jona, who teaches music to Seema and Monish.

My father is a devout Hindu, once professor of economics, a voracious reader, and very knowledgeable in Hindu scriptures. He now finds most comfort in the company of divine at the age of eighty. He acknowledges the presence of God in him, and he surrenders his mortal binary self to God. Most of his day, he remembers God and recites his name and derives incredible joy through this process. Hindus call this joy "Hari Ras" or divine intoxication. Whosoever communes with the Word, Shabd, or Naam will feel it enlivening effect. It is sweet and absorbing. Any description using words will fall short of describing its true reality. Far from being inebriating as the name suggests, it lifts one into a state of universal awareness, an incredible state of superconsciousness. All is simply known in this state. Once a quantum self tastes this sweet juice, it must have it forever.

> O Lord! Grant me the sweet elixir of Naam.
>
> Guru Arjan

> How unfortunate they are who do not get
> Hari Ras and are ever in the clutches of Death.
> A soul in contact with Hari Naam,
> Tastes that sweet elixir of life.
>
> Guru Ram Das

For me, the understanding of what it means to be in divine company did not come easy. As a young man, the name divinity or God was synonymous with laziness in my dictionary. This is despite my father and mother being deeply religious. I am not quite sure why. It may be because I was ambitious and very much influenced by the glitter of material wealth from the west. Growing up in India, I saw many baggers, many con artists who took, stole, or extorted money from unsuspecting simple people in the name of God. The sadhu or saint, according to my perspective, at that time, would be better of getting jobs than asking for alms for sustenance. I had no religious beliefs. If I could not see it and science could not confirm it, there was no God for me. The religion, according to my young mind, was for weaks who did not know any better.

Following my training and education in science and work experience, in industrial applied research pursuits, I maintained my views. A lingering feeling of discontent always pushed me to look for refinements in my outlooks as I searched for deeper meaning of life. I looked hard for deeper meanings of our existence using scientific, especially hard-core, materialistic views that were accepted in the scientific community. Even after searching for decades, I was not much closer to any significantly satisfying meaning to what was boiling inside of me as unanswered questions. I consoled myself by recognizing that the subject matter is immensely complex and involved many disciplines, and my answers would take time. I just needed to continue to dig deeper. In the mean time, I was drifting further and further away from any sense of divinity that I had as a child. My belief system rooted in classical science allowed only binary egoic self residing in my body and slowly divinity oozed out of my being. My binary self was perpetually engaged in binary computations directed to seek answers. My body started showing visible signs of revolt. I was becoming chronically unhappy. I was constantly sick, and my doctors had no cure for me. I knew something was seriously wrong inside. I knew, if I did not do anything extraordinary, I would die or suffer some crippling disability. I was on hectic reading spree of books covering self-help, psychology, social, and emotional and covering all sorts of scientific advances. I was on the hunt for anything that would show me respite. By chance, in an Indian gathering, in Artesia, California, someone tapped my shoulder and said,

"Here, take this DVD, it is free."

"What is it?" I asked.

"This is introduction CD to Maharajji." He is a spiritual master.

I gave it a quick glance and said, "No, thank you," as I had no time for such masters.

He looked deep in my eyes and said, "Why don't you try? You know you need it."

Against my own will, I took the material and heard the CD. Frankly speaking, I got nothing out of the first listen, but something in me persuaded to play that again. My rational mind still thought of its content as nonsense, but I listen to it again and again. For a week, my car stereo did not play anything but that CD. A week later, I joined the list of aspirant who sought "self-knowledge," the culmination of Maharaji's teachings. After almost one year of listening to literally hundreds of Maharaji's DVDs, I was given the gift of self-knowledge.

There is no doubt that just the process of receiving and practicing techniques of self-knowledge did what years of medicine could not do. I started to get better. I reduced my medicine intake significantly and was well on my way to be medicine free. Unfortunately, I made what I would call as perhaps the worst student of Majaraji. My quest to connect my scientific knowledge to ultimate meaning was so intense that I continued my search, now following Osho, another spiritual master from India. This state of bliss or enlightenment was the principal promise of both Maharaji and Osho's teachings. As I heard hundreds, if not thousands, of discourses from Bhagwan Osho, my clarity to achieve a state of bliss by the human body had intensified. My mind was receptive to achieving the bliss state of being.

Central tenet of teachings of both Maharajji and Osho lies in recognition of a God that resides inside of each of us. God is infinite with no beginning or end. How can it reside in our finite bodies? I was curious to how can one explain or at least be so sure of such bizarre entity with no background in quantum physics.

"It resides in the unmanifested realm. Our body is finite and subject to death and birth. Our body is a union where finite and infinite meets. We cannot see the infinite, but we can feel it," said Praveena, who initiated me to Majaraji's world and had become a good friend.

According to Maharajji, there are four key techniques of self-knowledge. These physical exercises aid your attention to be focused within of yourself. Maharaji would call these as equivalent to a mirror which shows you, your true self. The four physical exercises are described in Geeta, Hindu holy book that means "song of life." Here God Krishna is imparting this ultimate knowledge to Arjuna in the form of Rāja Yoga or royal wisdom. The physical exercises by themselves are not sufficient to truly go within where Maharajji wants you to go. The path to get there essentially involves dissolution of binary self in quantum self and, therefore, knowing one's true self. For this, one must be constantly listening to Maharajji. Each of his discourse is like cleaning your internal informational space of egoic influence. The modern living contaminates one's informational space with the significance of egoic entity, and its utter and complete dissolution is essential. Once it happens, subjective knowing of the cosmic mind takes place. Only then, one is introduced to one's true self, i.e., the quantum *self*. Maharaji calls his prime

When I say thank you or love you or express my sincere gratitude toward HIM, he answers in form of a feeling I can merge into this and surrender and enjoy the ultimate . . .

—Anonymous

goal as introducing you to your true self. Once in the company of divine, amazing sense of contentment results. Additionally, by being receptive to communication from divinity or the cosmic mind leads to "knowing" without reason, just know with clarity or light.

The state of self, completely under the surrender state to the nonlocal cosmic mind, is our true self or quantum self, as opposed to the other nature of self, the binary self. From a computational perspective, this is a state that is the image of the superpositional state of quantum particles. In the superpositional state, as I described earlier, each of the quantum entity loses its self, it combines with others to engage in true parallel computations that make the community of these entities act as one.

Imagine, all the human population on earth acting as if there is no egoic self. They exist for others. A combined entity of such quantum selves will mirror the divine at the quantum level. Widespread altruistic nature in living beings arises from this nature of quantum self.

Unlike the egoic self, the quantum self has no doubts about the existence of divine. It knows that divine exists, and it desires to seek its company all the time. It is highly tuned to the wave of guidance coming from divine, and it follows it with undivided trust. It appreciates all there is and feels incredible gratitude for all that is done by the rest to support its being in this world. All actions are attributed to divine as it believes that nothing and absolutely nothing can be accomplished without the grand will. It constantly communicates with divine and accepts all potentialities with grace and gratitude.

2. Quantum Oneness: Us, We, and Ours, My Quantum Self Is Selfless and Seeks Oneness

Being selfless is the true nature of myself. At the basic level, it represents a quantum computational entity, which as the processor of the information would like the outcome in favor of the rest or the whole but not the self.

How is this possible? It is possible, because the quantum self does not see separateness. It sees only oneness. By nature, this self is aware of the roots of

All the pain is in separation and all the bliss are in identification with the whole self (unity consciousness). Our Ego (a binary processor, brain) finds boundary, separates us from the rest, and splits rest into bits and pieces that can be used to make a picture of the universe. Real me is a quantum processor, it is a parallel processor, takes all information at the same time. It creates, shapeless, timeless reality where good, bad, positive, negative, wave, particle, win, loose; all dualities disappear. To access this reality, our binary processor (brain) must be in total acceptance mode, preferably in appreciation and gratitude to the ultimate self, the real self, in silence ready and willing to accept the infinite nature of *now*, the timeless nature of *now*, and the eternity called *now*.

—Anonymous

quantum processing that whole carries through superpositional state where all separateness coexist as one.

All its behaviors are based on this fundamental understanding. This is our true self. It is also the bottom-up understanding of our being. It is present in all of us. Saints have become saint by discovering the true nature of this quantum self in them. In the animal kingdom, the behavior leading to altruistic sacrifice of the individual for the survival of the species arises from this type of thought processing. Dennett provides an example of an altruistic society as he describes the Hutterite communities. These are colonies of so-called human bees where each member works for a communal good. What prompts a stranger to jump into a fire and save people he never met before? Clearly there is no time for weighing pros and cons. It happens because it is simply the nature of this self.

Besides humans, this nature of self is observed in countless of other living species. It is programmed in these species as instincts and leads to a variety of cooperative behaviors. Often this is detrimental to the individual but contributes to the survival of the species. It is more common in species with complex social structures. For example, vampire bats feed other members of their group, who have failed to feed themselves. Birds often help other birds, especially while breeding or protecting the nest from predators or help feed the young ones. Many animals alarm others in the group of the presence of predators despite the risk to their personal safety. Vervet monkeys are a good example of such behavior.

Many insects, for example, ants, wasps, bees, and termites are born as sterile workers. They devote their whole lives to care for the queen and the colony, a great example of true altruistic nature of their self.

The nature of my true self also has tendencies to center around "joy." Unlike the joy sought by egoic entity, this joy is without any side effects. The joy is realized by achieving the state of contentment. In other words, my quantum self is aware of its natural contentment state or self-satisfaction state. It acts in a way that it reaches this state to feel contentment or joy. My quantum self is truthful, empathetic, loving, compassionate, altruistic which stays in the company of divine.

Reason, nonreason
Wave, particle
Something, nothing
Perception, reality
Still, moving
Energetic, silence
Good, bad
God, evil
Awake, sleep
Living, nonliving
Life, death
And many more,
All dualities are essential
To make one *complete*.

—Anonymous

The self is needy. It needs energy to operate. It needs to reproduce. It needs to commune with the whole. The self depends on the rest or others in the whole for fulfillment of many of these desires and needs. The rest has a nature of change, and it constantly challenges myself to achieve fulfillment and joy. I feel joy when I am content. When my true needs are fulfilled, I am content; else, I seek to fulfill these needs. Depending on the need, I go within or outside (rest) to fulfill it.

I stay dissatisfied until my true needs are met. I must choose whether to rely on my egoic self of my inner, true quantum self to meet my needs. This needs to be a wise decision. While my egoic self is clever, crafty, game playing, my true quantum self only act according to its true nature. The informational entity, quantum self, dominated human behavior in our premodern lives. Now, it is the egoic self, which dominates our modern lives. Egoic self has been responsible for a large number of human comforts as well as sufferings that has marked collective human consciousness. As humans weigh out advantages and disadvantages of being in one of the two selves or a balance of the two, a new collective understanding is emerging leading to the resurgence of quantum self.

My true self has digital emotions, i.e., either they are or they are not. Yes or no. In its purity, love is digital in nature. It is there or not. Empathy, in the same way, is there or not. Truth also is there or not. The state of love is there because it exists, not because of a reason. Just like quantum particles, these states of quantum self leap from one state to the other without any transition in between. One arrives at these true digital emotions by connecting to one's true nature, not by the reasoning of the rational mind.

Reasoned love is ephemeral. My egoic self needs a reason to do something. It, usually, is its immediate or hidden welfare. If you help someone, and that person has not thanked you, and you are upset that your help was not even acknowledged with a "thank you," your help has not come from your true self. If you have loved someone, and there is a reason to do so, your love is temporary and will go away along with the reason. True self does not need a reason; it helps because that is its nature. It loves because that is its nature. It is what it is. Therefore, it is essential for me to know my true nature.

My true self just likes to be. It feels same inside out, in real Time, and provides immediate and truthful response. It likes to stay in present and

True divinity is where all dualities are accepted
in harmony. Reflecting a color of divinity means
accepting *all* including dualities with blessing.
Unity consciousness is the only real reality we have.
It is also the only way to true happiness.

—Anonymous

is content with what is and thankful and appreciative of all there is. It establishes a truthful connection with the whole and the rest. The roots of such behaviors of my true self lie in a well-known information principle called decoherence. It essentially means that you like to share whatever you have with your local and nonlocal neighbors. All entities in our universe follow this. It might involve establishment of entanglements to facilitate flow of nonlocal information between the self and the rest. The information contained in entanglements is so delicate that it cannot be manipulated and can only be stated truthfully. When you see a movie, you are eager to talk about it to others. You are decohering information according to this nature's principle of sharing information. When we reach home, after hard day of work, we are eager to narrate our day's experience to our spouse, we are operating per decoherence principle. All we see, feel, or experience can be attributed to information flow enabled by this principle of decoherence that embodies truthfulness. Photons from Sun, fall on my dark world, take measurements, and provide that information to my eyes, and make my world visible. The molecules of air that I inhale or exhale, as I breathe, are also busy in making such measurements and disseminating information to my senses through decoherence. All my muscles, all the cells, information molecules within me are also communicating using the same principle with utter truth underlying the information exchange. The whole universe is truthfully exchanging information, making measurements on each other, helping each other, and keeping track of each other's welfare all the time. It is like a lot of informational gifts are being exchanged. No gift is too small. No act is big or small. No entity is big or small. Whatever anyone can do and provide to others, it is accepted by all.

Exception might be the humans. Through the power of our rational mind, we have learned to lie, hide the truth or sugarcoat it to sound different than what it really is. When you lie, you are operating against this divine law. When you attempt to hide information, you are doing the same. If you are same inside out, truthful, you are according to this universal principle. Truth is the ultimate force behind quantum self. In his autobiography, The Story of My Experiments with Truth, Mohandas Karam Chand Gandhi developed the concept of soul force what he called "a quiet and irresistible pursuit of truth." Truth according to Gandhi had to be discovered experimentally in each situation. The true nature of myself is, in fact, for others. It is empathetic, kind, and appreciative of others. It is also an altruistic self with unselfish concern for the welfare of others: selflessness. It is always in a state of love with the whole and the rest.

Confession of an Ordained Minister
Gary Amirault, founder of Tentmaker Ministries.

Being an ordained minister for 20 years, the Lord has been able to teach me one primary principle: selflessness.

What is true, authentic, genuine, real, legitimate, and effective Christianity? Selfless service to the Savior.

But we must put this principle into practical reality by obeying God's Word in putting others before us, thinking more highly of others than we do of ourselves, and helping our fellow man make it along this journey toward home—heaven.

How do we do this? How have I done this? By secretly assisting the poor, the disabled, the needy, the orphans, the widows, the outcasts, the homeless children that fill our inner cities, and the less fortunate. Why? Because, it is more blessed to give than to receive.

One thing I have learned: this whole thing is not about me or you friend. It is all about Jesus Christ . . . and being a doer of the Word of God, and not only a hypocritical hearer.

Selflessness is the key to receiving the abundant life, favor, and blessings of God. So start today by taking up your cross, denying yourself [your will, your dreams, your desires, your ambitions], and follow the selfless suffering servant—Jesus Christ.

The Truth is not a religion, the correct denomination, the teachings of some great man or woman, a body of knowledge, the "right" book, a deep philosophy, the correct concept or a set of laws or governing principles. It can never be ascertained through the "scientific method," logic or reasoning. All these are instruments that fall far short of being able to fathom "the Truth." The Truth is a person who can only be known through a deep, intimate, personal relationship in covenant. Covenant is an exchange of lives. You will NEVER know the truth until "The Truth" becomes your life and your life becomes His. Truth requires absolute, total surrender to the Son of God. It is an exchange of life in the most intimate way in the universe. It is a marriage of souls. It is a bond that sets free. It is a paradox. And, when you enter this paradox, you will be free indeed, AND you will then also know True LOVE."

Empathy, compassion are the forces through which quantum self feels one with others. Empathy is that human ability, which allows us to imagine ourselves in another's position. Through empathy, we can understand how others think and feel. We can enter another person's emotional world and to respond with compassion. Empathy is essential for all healthy human relationships, social and personal peace. Having empathy requires paying attention to others rather than ourselves. It requires observation, active listening, and ability to focus and read nonverbal messages. To have empathy, we need to understand the true nature of our quantum self.

Empathy Waves, My Grandmother, Mother, and Wife

As a little child, I met my grandmother only a few times. She, however, left incredible impression on me. A frail, old, and weak lady with difficulties in walking. I used to be struck with the way she would speak and behave. For a long time, I could not understand why. Her demeanor, tone of voice and content all reflected incredible concern for whatever she saw. In almost incredibly humble manner she would blabber on, "Have you eaten? You look so thin. Are you taking care of yourself?" It was nonstop chatter, expressing incessant concern for others whether she could help or not. This was very odd to me as a child, but I loved it.

After I had gone past a few disciplinary relapse of my mother, I realized a remarkably similar behavior in her. I often would catch her in empathic waves of blabbering. Why don't you eat? Why are you not wearing enough clothes? Any thoughts that she would have would be a form of empathy or concerns for those involved. I was too young to understand or appreciate that mode of thinking. I simply thought that it was a normal style for older women to deal with others.

When Seema and I got married, we were young, and her style was different from my mother and grandmother. It was a relief in some ways. I had forgotten about the empathy waves, for many years, until recently, as Seema and I were walking together, I found her blabbering on, the same empathy waves. It took me a while to realize that Seema fell in the same category as my mother and grandmother. Now that I am sensitive to this empathy wave, I notice this all around. I catch both men and women to get on this empathetic train as a part of their normal thinking process. Although my unscientific opinion says more women get on this empathy train than men, but I could be wrong. Now, I think I understand why. True empathy makes us feel good. We latch on and adopt behaviors and habits that provide us that feeling of goodness. In Buddhist tradition, empathy and compassion are the two most significant traits that eradicate suffering from human existence. Both these traits are the nature of quantum self. It thinks all as one. Their pain and suffering becomes the pain and suffering of the self.

Love in action

September 19, 2009, was a special day for Raj and Grace. In a beautiful resort at gorgeous Santa Barbara, the setting was perfect. It was their wedding day. Grace was dressed in exquisite white whereas Raj looked debonair in exclusive Italian suit. Both were in fifties and without a partner for many years. A small group of close friends gathered to witness the ceremony. Raj first read his wows. Grace, overwhelmed with emotions, read hers in slight Korean accent. The wedding ceremony started.

> Dearly Beloved, we are gathered here today in the presence of these witnesses, to join Raj and Grace in matrimony, which is commended to be honorable among all men; and, therefore—is not by any—to be entered into unadvisedly or lightly but reverently, discreetly, advisedly, and solemnly. Into this holy estate these two persons present now come to be joined. If any person can show just cause why they may not be joined together, let them speak now or forever hold their peace.

Raj and Grace got married and moved to their Beverly Hills, home in Los Angeles. Raj has been a close family friend to us for many decades. His desire to find a soul mate has been obvious. Finally, that deep desire to find the soul mate is fulfilled. And he is content. True love is the hallmark of our quantum self. In a quantum bit, love and logic are combined. Love unites. It makes us feel one. We love because it makes us feel good. When we love someone we feel one with her. What will be the intensity of love when we love the whole existence? What does it feel like to be one with the whole? Our true quantum self has an intense desire to love, to be loved and to be with someone. It is the most fundamental engine that derives our quantum self. This, in fact, may be true for all form of life.

Story of Selfless Heroism
By Dan Abrams

I want to tell you about Private Stephen Tschiderer, a 20-year-old Army medic of the 265th Brigade Combat team. While on patrol, in Baghdad on July 2, insurgent snipers stalked the soldier by videotaping him from a nearby van. This footage would later prove Tschiderer's valor.

Within seconds of walking away from his Hummer, he was shot in the chest above the heart. Tschiderer goes down—but immediately pops back up, fires back, and runs back behind the Hummer. From behind the Hummer, he signals the snipers' hiding spot to his unit.

The 265th disabled the insurgents—including the sniper, who attempted to flee from the location on foot. Following a blood trail, U.S. soldiers located the wounded sniper and took him into custody.

While this is an amazing story on its own, after being shot by the sniper, Tschiderer acted like a true American by providing the man who had just tried to kill him with medical treatment. Tschiderer wrote in an email to his mom back at home: "Treating the man who shot me didn't really sink in until after. At the time, I just did my job and didn't think about it too much."

This is another example of the type of story the media needs to keep talking about—the selfless heroism of the American servicemen and women. Happily, Tschiderer is expected to make a full recovery, thanks to the body armor he was wearing.

The Hot Green Tea, River of Sweet Moments

In the beginning of this book, I conveyed that I really have no qualifications to write this book about eternal happiness except that I stayed in perpetually unhappy state for a long time. As I was finishing the book, another fact that matches the truth, started to surface. As I looked back, I realized that I had no plans of writing a book either. In fact, most content that has been written; I understood it well after it was written. It was as if this book was written through me, the title to the end. As I struggled with ideas, concepts throughout, the help automatically showed up. The clarity almost always came in the form of a feeling. This sounds so unscientific and almost something that one can make up, yet it is the cornerstone of this reality. It is the basic nature of quantum self. Here, the cosmic mind cocreates reality along with the rational and emotional mind of the self. The binary entity becomes an essential part of cosmic entity, the egoic self dissolves and concern for others become paramount. The whole cosmos acts as one, and all acts aim to produce an immediate or potential harmonious outcomes or "sweet moments" as described by Dr. Rekha Singhal, a professor and social activist, teaching in Jesus and Mary College, in Delhi, India. I have been living in it for some time now. Deepak Chopra, renowned author of several books, calls this phenomenon as synchrodestiny. I will illustrate this with one example that happened with me recently.

Seema, my wife, has been asking me to replace the carpet flooring in our house for a while, and I decided to see what Home Depot had to offer during my lunch hour. Coming out of Home Depot in Valencia, California, close to where I work, I was a little under the weather and felt like drinking something hot. Green tea popped in my head. It sounded like a great idea. Anyway, I thought green tea is normally available in Japanese restaurants, and I had no plans or time to visit one. My goal was to do comparison-shopping at Lows Department Store about five miles from Home Depot and get back to work. As I arrived at Lows store on bouquet canyon, I found myself getting on the parking lot, but somehow, I did not park, went out, and continued on to Bouquet canyon toward Valencia Boulevard. I was surprised at myself. How could I not go in after driving so many miles? For a moment, I did not know where I was going as I passed Valencia Boulevard heading south on Bouquet Canyon. I should have taken a right turn on Valencia Boulevard. Some frustration was setting in against myself as I was heading off to the route nearly opposite of where my work was and where I should be heading. I argued within my mind that there is a magic mountain park way that I can take to get back to my workplace. As that was going on in my mind and I was pretty sure of where to go, I see this flooring place on my right side. Immediately, I argued that I could check out this store while I had a few minutes to spare. I had to scramble into the side lane to gain entrance into the parking lot. As I entered the parking lot, there was a parking spot, right in front of this store, which I missed, and I wondered how could I miss such a convenient spot. I could have gone in and out of the store in no time. Anyway, there was more parking as I headed toward a parking structure right of me. I parked and got out. I had never been to this place. There was this plaza on the side. The flooring store was just a short walk away. I walked toward the flooring place. I saw this restaurant. The nameplate said the Tea Garden. I had never heard of a restaurant by that name, sounded kind of funny. There were a few people in. I went in thinking I may get a quick bite to eat as well. My turn came soon. I asked the woman if she had any sandwiches for lunch. She smiled and said, "Sorry we do not have any sandwiches." Surprised and amazed as I did see a few people sitting on the table, I asked, "So what do you sell?" "We only sell tea, here is the menu." First category on the menu was green tea, in fact, many varieties of them. I ordered a glass of hot green tea. It must have been only fifteen minutes since that impossible idea of having hot green tea entered my mind. The hot green tea I was holding and sipping.

According to Deepak Chopra, acausal nonlocal association can explain such instances. However, for Rekha Singhal, it is simply an act of cosmic whole, whose outcome is what she describes as "sweet moment." According to her, we have a choice of acting in such a way that each event results into sweet moment for all involved. "It is the call of our heart," she exclaimed. If all of our acts, even the tiniest one, follow such algorithm or goal, our lives will become rivers of "sweet moments." This would be a sharp contrast to a life, which comprises of moments of acts dedicated to the benefit of self and eventually deteriorating to acts of diminishing returns. She believes that it is the true nature of nonlocal, cosmic mind, or God. The God, according to her, is simply a grand collection of such sweet moments. "The cosmic mind is omniscient, omnipresent, and omnipotent. It hears all, and all its acts are such that they intend to produce a river of sweet moments for all. The fact that you got the hot tea you desired, from pure intentions, was simply one such act of grand cosmic mind that produced a "sweet moment for you," she described to me in her recent visit to Los Angeles. Since she happens to be my older sister, I was carefully taking notes. The sheer nature, number, and clarity of such events have been such that I cannot help but being baffled by the presence of this entity that almost appears to hear me and make universe react to what I do or simply think or desire. Therefore, as a person with roots in hard-core science, I am not sure how this book came about or who wrote this book through me. As a spiritual, informational entity, I know that the real author is not me; just like everything else, it has come from the same one source, a matching truth that only I know, for sure.

3. My Quantum Self Is Many Bodies One Mind

My quantum self plays match game with divine or whole, the mega quantum processor that runs this universe. Just like the whole, my quantum self creates one mind state with others it comes in contact with. Cooperation, togetherness, sharing and the common good are the objectives. Such behavior is prevalent in many social animals. Schools of fish act in concert as one mind. They almost act as one. If a predator approaches, the group not only senses but takes action. Each individual does not take separate action, but all act seamlessly as a colossal group. They can spontaneously form a large ball or scatter once the danger is sensed. Is there a "sixth sense" that guides their movements and coordinate their behavior?

Flocks of bird display amazing coordination as if they were all directed by one mind. The colonies of bees act in concert as one being. Being together, working together, solving problems together is the hallmark of human society. Early humans, living in smaller groups gained superiority over other species through acting as one mind. We have recognized that we may be many in body, but if become one or coordinated in mind, we can achieve more compared to if we acted separately as individuals with separate minds. King Chou of Yin with seven hundred thousand soldiers battled against King Wu with his mere eight hundred men. Yet King Chou's army lost because they lacked unity. King Wu's men won because of perfect unity.

Corporations, much like many other organizations today, recognize the value of teamwork. The monetary incentives are very popular and effective especially to an egoic mind, aiming to achieve results. They, in fact, constitute the cornerstone of capitalistic societies that have dominated our world as economic powers. Volunteers often volunteer as they are concerned about the needs and the well-being of others as well as the desire to make changes and improve the whole or the community. So many of these volunteers are driven by their quantum self. Volunteer organizations often rely on inspiration of members for their effectiveness rather than the monetary incentives. Even though inspiration is a powerful force to create one mind in our modern society, operating a volunteer organization is full of challenges. One can compare these organizations as the altruistic bacteria colonies in predominantly mean bacteria societies as described by Robert Austin. How do they survive? Can they thrive? Some insight comes from the open source communities.

Bottom-up Growth, Story of Linux

The community of Linux users and developers is held together by pride and the thrill of working toward a common goal to provide an alternative to Windows.

He was only twenty-one years old living at home in Finland; Linus Torvalds created Linux operating system. He released Linux as open source software in 1991. He offered it free to the world and made the source code available to anyone who wanted to alter it—as long as the tinkerer was willing to make the new additions available to the public as well.

Many used Linux because it was free. The product was eventually embraced by the geek community and penetrated the mainstream, running servers and other hardware. It represented an alternative to windows, but no one could foresee a rosy future for this product, especially as a competitor to Windows. Microsoft had lots of resources while Torvalds had virtually none in comparison.

Today Linux is a worthy opponent to Windows. Business Week/Online rated Linux wealthier than Microsoft because "Torvalds can muster more creativity from his far-flung rank and file volunteers than Bill Gates could from his corporate monolith."

Without any doubts, in today's society money is a great motivator. Microsoft has lots of it and uses effectively to motivate its programmers. Torvalds, with far less financial resources, has a bigger team, potentially millions who use Linux and continue to tinker with it.

Torvalds has another advantage. He is not controlling the growth of Linux. It is the innovation, creativity, and need of the mass of folks underneath that drive its growth. Such bottom-up system can be more innovative than the hierarchical system developed in corporations such as Microsoft.

The Open Source movement is an amazing example of the organizations that follow and foster one mind and many bodies nature of quantum self. Such organizations are attracting and engaging a new generation of creative, innovative people. There is a free-flow exchange of ideas and products within the community. There is the challenge of creating a better product/ service. There are opportunities for both individual and team creativity and innovation. There is the opportunity to share in a common vision with likeminded people. There is an environment in which participants can work as equal partners with some of the best and brightest in the field. There are opportunities for dialogue and community. Though much of the activity takes place as part of a vast online community, many local communities have monthly face-to-face meetings, creating technical support groups and open source communities that foster the exchange of knowledge and opportunities to build skills and connections.

The Open Source community thrives on individual contribution to the collective whole, innovation, creativity, and the exchange of knowledge. Participants are passionate about the work and the product. There are no external recognition systems. Reward is the intrinsic value of being a part of this unique community that reflects personal values and beliefs. There are no officers or committees. There is the flexibility to move in out of the project as personal time requires. There is great self-satisfaction in being a contributing member of this worldwide community dedicated to the development and distribution of free software. These volunteers are about making a significant change through collective action based on individual contributions.

4. My Quantum Self Is a Doer Self

As described earlier, all entities within the whole and the whole or the cosmic mind by itself is taking constant measurements on anything and everything as well as decohering or broadcasting the gathered information to all, all the time. It may be one of the most fundamental natures of the informational universe. How does that work? It is simple. At any moment, the whole is filled with photons, cosmic rays matter particles. All the photons, cosmic rays, neutrinos, atoms, molecules, and larger entities are bumping into each other, and that is all it takes to make measurement on each other.

If there is something that you think is yours, the whole and others within the whole wants to know what that is. If you wanted to hide something from the whole, it would not be an easy task. The whole is omnipotent and omniscient. From smallest to the largest are all observed and kept track of through constant measurements. All inhabitants appear to decoher the information they gather as soon as they gather. It is quite like broadcasting what they know truthfully when they know. Well, we know this may not be entirely true with humans. We do keep some secrets, at least we try to.

These eavesdropping of the whole on everything and everyone is the key for the formation of all visible structures, including us, in our universe. You may ask how does the whole make observations in the deep space with utter vacuum? Even in the vacuum, particles wink in and out of existence and the collisions results in making the measurements. All the interactions and measurements that this whole performs on itself and its entities appear to govern it's, own internal structure and its nature from the bottom-up perspective. Considering this need of the whole to sense and create its own internal structure, it may not be far-fetched to believe that entities produced by the whole act as a sensor or even cocreator in this whole process of existence and creation. We may also call them eyes and ears of the whole or simply helpers.

From this perspective, the whole may be busy in building superior sensor technology that provides it better description of itself as well as cocreator technology that would help it create the existence. Life may be one such sensor and cocreator technology for the whole to make that measurement and perform a task to cocreate forms of it. Humans with a sophisticated set of minds namely emotional and rational may be the most complicated sensor and cocreator technology the whole has ever produced. The connection of these minds with the cosmic mind may complete this communication link.

A relevant analogy may be human efforts to understand a far away planet such as Mars. Recently National Aeronautic and Space Administration, NASA's

Mars rover Spirit marked six years of unprecedented science exploration and inspiration. The Mars rover Spirit successfully landed on the Red Planet on January 3, 2004, and began missions intended to last for three months but which has lasted six Earth years, or 3.2 Mars years. It was packed with instrumentation that enables us on Earth to know the local climate and conditions on Mars. During this time, Spirit found evidence of steamy and violent environment on ancient Mars that was quite different from what opportunity, another sister Rover, operating and exploring another site on the planet, found. Someday Rovers may become intelligent enough to cocreate Mar's landscape along with us here on Earth.

We may be one of the many rovers of the cosmic mind. This just might make a great theme for a Hollywood science fiction or may be a science fact, but my quantum self sees itself as a cocreator of this existence, as a humble servant operating under the guidance of the whole.

My quantum self has binary self hidden inside. It is capable of all binary mind can do. With one exception, it simply acts and surrenders all the results to the whole. All its actions benefit the whole not the self. My true self is conscious of not only my physical self and the physical rest but also the whole. It is an observer and interpreter of the whole and participates in creating the whole by being a doer. It is highly tuned to communication from the whole and the rest. It appreciates the true meaning of the phrase "quantum of action." It is a doer. Any type of service or work especially in the service of others will do—action for action sake, not governed by the results. It simply does and leaves the results to the whole as a part of the grand quantum distribution system operated by the whole. In other words, it performs the action as a part of the command from the whole. The whole is always for common good. The quantum self is to serve others. It naturally gravitates toward group dynamics that leads to participation by all and for all. The common good is at the heart of all individuals and communal actions. Greed, avarice, self-interest are all seen as undesirable traits.

Public Service, Leadership, and Management for Common Good

Humans driven by the quantum self are good leaders as well as followers. As leaders, they project a leadership style that has a sense of selfless, public service for common good. Their style of leadership influences others to accomplish shared objectives and make the organization more unified and coherent. They believe the basis of good leadership is having an honorable character and providing selfless service to your organization.

People focused on their quantum self are effective communicators as their honesty and transparency leads to winning organizational trust and confidence. Most people want to be guided by those they respect, trust with a clear sense of direction. Such leaders, in general, earn their trust and get many followers. Trust and confidence in top leadership is the single most factors that affects the level of satisfaction in an organization or in a group. To gain respect, they must be ethical. Leaders tuned to their quantum self are ethical by nature. They also possess a sense of direction, which is achieved by having a strong vision of the future.

The quantum self, as a leader, facilitates volunteer involvement with in a group or an organization. They promote shared vision and leadership. A shared leadership involving leaders at all levels starting with the board members, the CEOs and the middle management, all working together as one mind. It is also one of the most effective models for encouraging volunteer involvement within organizations.

The quantum self as a leader is a terrific coach. Coaching is less about teaching and more about helping people to learn. As a coach, the quantum self listens, supports, encourage, and empower others. He guides, leads, maximizes volunteer resources by helping individuals master their work, know-how and skills. He believes in unlocking others inner potential and maximizes their own performance.

As a coach, quantum self fosters one-to-one relationships. He invests time. He often helps others discover the answers to their own question, rather than give them his brand of solution. It is often easy to give a quick answer instead of taking the time to coach and discover their answer. However, for quantum self, coaching is like the ancient Chinese proverb, "Give a man a fish and he will eat for a day. Teach a man to fish and he will eat for the rest of his life." The story of Muhammad Yunus, winner of 2006 Nobel Peace Prize, is truly inspirational in making such proverb a reality.

Nobel Peace Prize Winner, Muhammad Yunus

Muhammad Yunus founded "the Grameen Bank" in 1976. At that time he was just a young economics professor at Bangladesh's University of Chittagong. "Grameen Bank" means "rural bank" in Bengali. Yunus lent to forty-two women in the village of Jobra to finance their small businesses of making bamboo furniture. The loans were small, equivalent of an average of U.S. $27 from his personal wealth. Yunus's strategy was to do the opposite as conventional banks would normally do. The banks normally lent to the rich; he lent to the poor. The banks normally lent to men, he lent to women. Banks normally required collateral; his loans were collateral free. Banks required a lot of paperwork, his loans required none. Unlike most of the bankers, he was friendly and trusted the women.

Since that humble beginning, his bank has made an estimated $5.7 billion in loans. It has a repayment rate in excess of 98 percent, which is highest among any banking institutions. The loans enable millions of poor people in Bangladesh to start and run their own businesses, lifting most of them out of poverty.

Muhammad Yunus was awarded the 1.4-million U.S. dollar, 2006 Nobel Peace Prize for pioneering a new category of banking based on the true altruistic nature of his quantum self. After receiving the news of the important award, Dr. Yunus announced that a part of his $1.4 million award money will make low-cost, high-nutrition food to be available for the poor while the rest would fund eye hospital for the poor Bangladeshis.

Self-assembled, Self-Powered, Computational Self, and Communities of Selves

Most scientists believe that the big bang took place around 14 billion years ago. What about prior to that moment in time when our universe began? Science does not have much to say about the time before the big bang. However, there is no reason not to believe that the emptiness or unmanifest that existed after existed before the big bang as well. The birth of our universe was actually the birth of the grand quantum-computing platform that we are all plugged into.

Ever since its birth, the computational entities have been evolving in ever so complex shapes and performances. Starting with quantum computational entities such as photons, subatomic particles such as electrons, photons, atoms, molecules leading to evolution of a living cell and finally today complex life based on multicellular organisms that act as a single computational entity. One common performance standards for these entire computational elements appears to be their ability to perform computation according to a qubit algorithm. It appears that devices undergo a perfection processes through trial and error until a final selection of a new generation of computing element is made by the cosmic mind.

From its basic computational units and patterns, it took millions of years of trial and error of chemical evolution before stage was set for first biological computational element to evolve. It took about a billion year for living cell to emerge. The experimental trials and error continued with a single-cell shape, nature, and performance until the cosmic mind chose certain type of cells to make larger computational elements.

It took another three to four billion years before the multicellular organism evolved. Here, for the cell to work, the informational circuitry had to be such that it can fundamentally operate in the image of a qubit. For last 12 billion years or so, the cosmic mind has been experimenting with the wide variety of multicellular computational elements, which is multicellular life in its all kind of variety of shapes and forms.

We get so used to seeing ourselves as one entity that we forget that each one of us is a community of trillions of living cells. Each of the living cells in its own right a sophisticated computational element, a living individual, a sentient being that has its own life, functions, and interacts with other cells very much like a busy city of trillions of individual cells.

Humans, for now, appear to have won the game of survival through mastery of the art of binary computing but still cannot ignore their roots of quantum computing to achieve harmony with the grand quantum computing performed at the deepest level of the cosmic mind. With two different nature of computing systems housing in one physical body, human device has discovered the meaning of "suffering," elimination of which continues to be primary occupation of human society, the holy grail of human existence that is achieving eternal joy.

Myself represents an entity or organism that is different from the constituent cells. It has the powers over the community of constituent cells at its command. It can make the whole organism move. It has at its disposal power and energy of all the constituent cells. Each cell is also a computational self in its own right and plays its part to simulate the information space that myself wishes to create. Each cell surrenders to the emergent self. Just like a leader, myself may create an environment of appreciation or gratitude toward its little solders or cells within me or create an environment where they are forced to work hard, create conflict, live in molecules of emotions created by negative emotional, informational fields. What myself understands and does clearly plays a critical role in emotional health of my being. Each human self, just like each cell in our body, makes a computational element that participates in a larger organization. Its happiness and well-being is affected by the type and the nature of the organizations it creates or a part of.

Will we be able to create other larger organization such as family, society, corporations, cities, and colonies that could run on quantum computational principles and would stay in harmony with the grand quantum computation carried out in the cosmic mind? It all depends on the evolutionary nature of human self. At the moment, it appears that our grand universal system the whole is betting on human self to be the building block of the next computational element that would lead to even larger computational elements. How good is this bet? Which self will win? The binary or the quantum self? As a cosmic mind, I would bet on the quantum self. Why? Because this mode of computation offers incredible computational efficiencies. What will really happen? Only time will tell.

———ⱳⱳ◦◦℮ⱦ◦◦ⱦ℮◦◦ⱳⱳ———

Chapter 5

Nonlocal Mind Unity
Consciousness Enlightenment

> *Non local self resides within. It is the soul and pure consciousness. In concert with local self it forms the self.*
>
> **—Anonymous**

It may not be a great surprise, to most, that we are incorrigible seekers of happiness, bliss, and joy. Longer and deeper the feeling the better it is. Now the science confirms that our brains are wired to feel good. Marketers have known this for long. It is no accident that our brains are constantly bombarded with messages that tempt us with products, services, or styles that would make us feel good. Many marketing slogans promise even nirvana by simply trying whatever they are selling: a pair of new shoes or new car or just about anything else. As good as we feel when we eat certain food, dress in certain clothes, have sex, indulge in romance, throw a great party, or engage in a great conversation, our capacity to feel good diminishes if we repeatedly overuse one or more of these activities to get pleasure. We, therefore, have to come to terms with the fact that we do not have unending appetite for the materials contents. This is a part of a well-known principle in economics called the law of diminishing returns. Most of our materials wishes are subject to this law. This simply means that our enjoyment drawn from an activity or object lessens with repeated use. No wonder we are constantly looking for novelties, new things to do to have fun. It does provide a formidable challenge to artists, creators, and corporations to come up with new products and services that can keep our modern society entertained. Over time, there is a limit to the joy we drive from novelties as well until the law of diminishing returns takes hold. As we reach old age, our faculties get older, and our mind gets saturated with acts of dwindling returns; our search for that timeless joy that our minds are wired for intensifies.

Is there anything that transcends this fate? According to Ayurvedic masters, deep feeling of contentment or enlightenment is one action that defies this principle of diminishing returns. According to them, at a deep subconscious level, this is what we are all searching.

What Is Enlightenment?

In its simplicity, enlightenment is the knowledge of the supreme truth. What is that ultimate truth, and why is it different from most other truths? I'm sitting on a chair, its color is black. It is the truth. However, it does not give me any happiness. The ultimate truth is about knowing the true nature of yourself and the universe that you are a part. Once you understand this, and your perspective reflects that knowing, unending bliss pours out of the core of your being. Once you know the supreme truth, no doubt remains,

Inside of me, there exists a sea of potentialities, formless, timeless. It constantly provides me answers for my mind to accept. The sea of potentiality is beyond space time and hence exists everywhere all the time.

—Anonymous

timeless, oneness, pure love, and extreme empathy feelings will fill the heart and mind of the knower. Gratitude and appreciation will reign supreme. Your heart will not stop thanking the supreme God for a chance to live this life. The feeling is intense, sweet, and so complete there are no descriptions that can do justice to its truth. No other truth can do all that.

So how can I know that truth? Well, it turns out that all of us have that truth in us, but most of us are not awakened to it. Enlightenment is awakening or awareness to this truth that lives within us. Enlightened souls are awakened to their true self. They are also awakened to the true nature of the rest and the whole. They are also awakened to the separation of self with the rest. They know giving up the self and merging it with the whole is an intensely pleasurable transition. Keeping self as a separate entity is where all pain and suffering lives. Since our modern society still fundamentally relies on separation of ourselves, enlightenment continues to be the holy grail of humankind achievement.

Yogis, Saints, and Enlightened, Masters of Time

Over the last several centuries of evolution of modern humans, the history is filled with examples and stories of human beings that we have called saints, yogis, spiritual masters that have claimed to have achieved enlightenment. Mahavir, Buddha, Mohammad, Krishna, Ram, and Jesus—some of the widely recognized masters of time founded some of the most widely followed religious and spiritual traditions in our societies. All have agreed that enlightenment is the experience of our inner being. To experience and understand enlightenment, we must probe deep into the nature of our inner informational self.

Your Inner Informational World, Your Attention and Intentions

The Blue Brain project was launched in 2005. It is, by far, the most ambitious brain simulation effort ever undertaken. Many computer simulations, in the past, tried to code action of the nervous systems in different animals. The Blue Brain project used a different approach. It reverse-engineered mammal brains, from real laboratory data, to develop computer models, that could mimic the behavior of the combined network of rational and emotional

At deeper level of self-realization, I'm a collection of quantum and binary computers that simultaneously process information in parallel in local and nonlocal domain. When I merge with this reality, I see myself as a silent witness to myriad of processes happening all in the same moment of time called *now*, no boundaries between me and others, a seamless existence, all coexisting into one. I feel love, empathy, and profound connectedness and fearlessness.

—Anonymous

mind. At the end of the first phase, the researchers had modeled the neocortical column or the neocortex, a critical part of our brain responsible for higher brain functions and thought.

"Thing about the neocortical column is that you can think of it as an isolated processor. It is very much the same from mouse to man—it gets a bit larger a bit wider in humans, but the circuit diagram is very similar," Henry Markram, leader of the Blue Brain project, told BBC News.

Henry Markram and his team hope to create a realistic digital 3-D model of the entire human brain within the next ten years. Mathematics underling the modeling of the brain does simplify things, but the sheer numbers of neural connections involved require a large computing time to make any realistic computations and simulations.

Markram and his team, for the first time, are able to shed light on how the brain and connected informational networks work. According to Markram, "The holy grail for neuroscience is to understand the design of the neocortical column. It will help us understand not just the brain, but perhaps physical reality. Understanding the structures that make it up is difficult, because beyond just cataloging the parts, you have to figure out how they work—and then build sensible digital models."

What Markram is trying to get is the glimpse of our inner informational world? He is attempting to understand the nature and characteristics of informational entities that live in our inner world. Through detailed computer simulation, Markram and his team finds that brain forms what he calls as "electrical objects" that are the direct result of neural activities in response to a mental process such as thinking a thought or viewing a physical object. Inside the mind, the physical universe is represented by a digital informational universe comprising of these electrical informational structures.

Even though modern science is only barely able to scratch the surface of this world through objective methods, ancient Ayurvedic seers, with their subjective mode of investigations, have a lot to say about this world. In fact, to them, knowing this world was paramount to achieving a sound, emotional state of being, happiness, and even enlightenment. Essentially, everything we experience as physical reality has a counterpart in our inner

Whatever you attend to becomes your truth. In a parallel distributed information network system, many parallel realities operate at the same time. I am pressured to choose, and choose wisely, to pick my reality. With all these distractions and fears I need to pick what gives me most *joy* and contentment.

—Anonymous

informational world, and the two worlds are intricately connected. Our physical or outer self lives in the outer physical world where as the most prominent informational entity that lives in our inner world is our inner or informational self. It stays with the rest, as another informational entity that represents everything besides the self. The two together forms the whole, the informational entity that is our own informational universe present with in each of us. This is our subjective universe. Here our opinion and understanding is paramount. Knowing the true nature of this inner world was critical to understanding the true nature of us, according to these seers.

Ancient Ayurvedic seers assigned the greatest significance to the purity and quality of our inner world in ones quest to achieve divinity. The word purity here means the truth, love, and compassion, the characteristics of the whole or the cosmic mind. The word yoga essentially means union. When our inner self unites with the inner whole, this union is called enlightenment.

Most of us are quite aware of our inner world and spend a lot of time living in it. It is the world of our thoughts, feelings, sensations, sounds that our inner world produces. Feelings are, by far, the most prominent mode of communication in this world. However, we do not pay enough attention to all the feelings. Sensations and feelings come and go, but we keep on going, doing our thing, often busy in our thought world. Unless our body forces us to attend to our feelings, we do not. This describes most of us. We are so used to seeing our reality in the form of our physical universe external to us and physical entities interacting in this external world that we often would deny the presence of this inner world of feelings as we cannot quite place a finger on its whereabout. This is the legacy of centuries of perceptual conditioning, based on materialism, a disease that most of us suffer from.

We rarely volunteer to witness our feelings silently. According to ancient seers, this just shows that we have not understood the significance of communication from our inner world. According to this wisdom, accepting and understanding communication from our inner world has a significant impact on our emotional state of being. Every thought counts, every sensation counts, and even every thought that is not even a thought matters. Even the intention proceeding thought counts. We are the architect of our inner world. Each memory, each experience makes a view of this world. We not only hold it within us but live in it as well, most of the time. We

Divinity is the truth that can be felt, it is instantaneous. Mind is mostly the perception that takes time to play out. I always give priority to the *truth* over perception. When I accept the truth in my life, I feel *joy*.

—Anonymous

come back to the hypothesis of WYSWYG or what you see is what you get. In light of the experience of the inner world, the two postulates of this hypothesis can be restated. First, the reality is what you feel or experience in your inner world. Second, there is no reality beyond this reality. This is exactly what ancient seeker named as our true reality.

Ancient seekers understood the significance of preserving the purity of our inner world, especially attentions and intentions leading to thoughts. Every thought, every feeling needed to be treated like we should treat a human being preferably with truth, empathy, trust, and acceptance.

Our inner world is the bottom up view of our informational reality. The reason it is called inner world is that it is formed out of the information comprising experience of our inner feelings stored in the form of memories. It constitutes an incredibly complex information network that is home to many informational entities. It, in fact, is an informational image of all we see, experience, and understand in the physical world. In this computational universe, there is no space and time. Everything is interconnected at local and nonlocal level by the interaction of three minds: rational, emotional, and the cosmic that constitutes our inner self. The space and time is created for us from information mediated by our attentions and intentions. Our attention brings some sense of rationality. It picks and attends to what is most meaningful to us. Our attention, one may say, is the observer within ourselves. Its significance rests in prioritizing what it chooses to observer or attends. Our intentions are what we intend to act next after whatever we are doing now or series of next acts following the ones going on now. This means that intention is critical to our next thoughts. Ancient seers considered these two elements to be critical constituents of the inner world.

In my inner world, at the most fundamental level, there is recognition of a divine like entity that is timeless, spaceless, and energyless. It is digital divine or the informational whole. It is a grand cosmic mind, with mega computing core, where every reality plugs into. It always was, is and it always will be. Our visible universe began with the formation of simple qubit devices soon after the big bang. These formed simple informational patterns which in turn, evolved to more complex informational entities. These became atoms, molecules and other larger particles. Eventually, even more complex informational entities such as living cells evolved.

There is local information that is space and time based and then there is non local information that requires no space and time. God resides here and is attached to all and is available to all in a seamless unity at this most fundamental level of our existence. Any efforts to maintain that unity (information level) lead to pleasure and any efforts that lead to break that unity such as *ego*-based reality leads to pain and sufferings.

—Anonymous

All living cells were formed in the image of qubit. Living cells evolved to form a variety of more complex living systems including humans with the inner self. From a bottom-up point of view, my inner self is an informational entity, which is also rooted in this reality. My inner world is a world of information. At birth, I have a blank slate of consciousness mapping my inner world. The story of my journey through physical space and time, my emotional story, will be written on this slate of consciousness. The main writer, producer, and director of the drama called me is the cosmic mind. I am the cowriter, coproducer, and codirector of this story. It involves much emotional drama, some achievements, some failures, and perhaps a lot of learning.

Our birth, in this material world, is like admission to a university, according to ancient Ayurvedic seers. Starting from humble beginnings, our inner world slowly houses more complex, informational entities, related to our learning and experiences of our physical world. Our inner world meticulously stores memories and experiences we have had in the form of informational files that could be opened or closed. Who is a part of my inner world, and how do I treat that entity with in me is key teaching of yoga practice. True meaning of yoga is union. It addresses how do I unite all that is living within me? Do I treat all within me with respect, honor, kindness, empathy, and love? Or do I become an egotistic ruler of my inner world, angry, frustrated, overpowering, and taking advantages of entities within me? Ancient Ayurvedic seers went to the extent and believed that if you have committed a rape or murder in your thoughts, it is no different that committing the same in the real life. Of course, if our legal system adopted such logic, most of us would be criminals.

If your inner world is engaged in conflict, reasoning, power struggle, greed, jealousy, your emotional state will be of unrest. These negative emotions will become roadblocks for you to achieving enlightenment. Your opinion or belief helps create the nature and the type of informational entities that are created in your inner world. Truth is all-important ingredients. Understanding and assumptions underlying the belief or opinion are also critical. It is clear from Markram's brain simulations that our mind only needs some data to create an inner reality for us. It adds a great deal of simulated reality, based on its experience and calibration to give us impression of a reality that we often believe is real. Mostly it does a superb job in doing so.

Best reason to do something is No reason.

—Anonymous

However, sometimes when these assumptions are forgotten, we often create incomplete image in our inner world. For example, when we create an image of another human being, our mind takes partial data about that individual and adds simulated information based on assumptions, experiences, and calibrations relevant to that person. However, we often forget that he or she is as complex as we are, and our inner informational images are based on assumptions that may not be true. We also tend to forget that he or she creates an image of us in his or her mind just as if we created his or her image in our mind. We often forget to consider that he or she also has the complete divinity or whole living in his or her inner world just as it does in our case. More often than not, we jump to conclusions about others based on our own perceptual beliefs.

Purification of your inner world is the essential element to achieving spiritual growth. In our inner world, each of ourselves is connected to others but is also connected to the whole or divine. Ancient seers advocated the link to divine to be lot more significant than the direct link to others. They favored a degree of detachment to any link that does not route through divine. The direct link would normally be corroded due to self-driven interests. This link should not get the importance unless it is cleanse to the purity of the whole.

In observing the inner world, the quality of intentionality is critical. If the observation is made with intentionality, such as of prayer quality or humility or surrender; then, information flows from divine or the cosmic mind to the rational mind. Else, it flows from rational mind to the body or emotional mind, the later needs to be avoided, for any spiritual uplifting especially, if it is rooted for self-interest. Rational mind depends on binary information processing. Reason is at the core of this entity. Rational mind or reason lives within the emotional mind.

The emotional mind is rooted in quantum information processing, which is inherently uncertain, but is intimately connected to reason through rational mind. Both local minds viz. rational and emotional mind arises out of randomness. One may say both these minds live in randomness or nonreason. Ultimately, both reason and nonreason merge into silence or emptiness or oneness, a nonmaterial superpositional state of existence that provides a quantum-computing platform for all that is visible or invisible. The nonreason is as important as the reason, if not more.

When your conscious mind or attention is focused on nonlocal *self* (or quantum *self* or real I) with full intent and accepts in earnest answers provided by the real I in form of feelings . . . enlightenment results.

—Anonymous

The ancient seers described enlightenment as the experience of this oneness or whole through deep emotional state of Samadhi. To gain entry into this portal, self would have to be like the whole. Truth and nothing but the truth prevails in this state. Empathy, unconditional love, true oneness of mind, and heart prevail here. In this state of self, all doubts disappear; reason is superseded by love and incredible sense of contentment and completeness results. The time vanishes. The self merges with whole in complete humility and surrender. The whole universe acts as one. It is the most common experience of Samadhi described by most meditator.

Meditation, Samadhi, and Kundilini Awakening

Meditation, Samadhi, and Kundilini all describe processes that fine-tune the state of our attention. Even though we could appear quiet from outside, in our inner world, we are usually engaged in multiple, parallel thoughts or activities. The attention picks one of these activities at a moment of time. If you are like most humans, your attention will jump, almost like a monkey, from one issue or thought to another. The ancient seers realized that concentration of the attention was a critical element to explore realities of the inner world. A key element that ancient seers realized was to achieve the state of focused attention.

Where the attention should be focused? It should be focused on the whole or the cosmic mind or divinity. Since our self is a play of three minds namely rational, emotional, and the cosmic, our attention is constantly trying to choose one of the three. In our modern existence, for most of us, our attention is usually glued to our egoic entity. Removing attention from egoic binary thinking and tuning it to our true quantum self and accepting the wave of guidance through sensing and feeling, ancient Ayurvedic seekers found the path to bliss. This became possible only when the attention could be focused and centered on present.

As the attention moved from a physical space-time reality and turned to a space less, timeless, God-like reality, the seers claimed that all was knowable in an instant. All motions were possible without movement, everything that was knowable was known in an instant, and a grand blissful feeling abounded the whole body. Hindus describe this experience as Sat-Chit-Ananda, which means truthful consciousness of divinity leading to bliss. The name is a

Dhayan is attention. Here intensity matters. More intense is the dhayan, greater the stillness, greater the connection to nonlocal self.

—Anonymous

kind of the key to reach this ultimate bliss. It also describes the most basic informational, computational nature of our true self or quantum self. It is always in tune with the communication from the whole. The computation performed by the whole or divinity produces results which are communicated to the self through feelings. The abilities to experience such feelings by the self are critical. When such experience, by a self, is accepted with purity or truth, the bliss is felt. The deep and extended feeling of such experience was so extraordinary that even millions of books, sermons, sat-sangs, songs that tried to describe it failed to do justice. It was called the feeling of divinity and a know-how to get to such feeling was the path to enlightenment. Therefore, one may say in our modern language, the enlightenment is nothing short of finding entry into the grand quantum computing system; the informational whole or simply to be able to tune in to the nonlocal communication from and within the whole. A few actually did.

Most recognizable and publicized are the experiences of Buddha after enlightenment. These experiences were so bizarre that a rational brain would have a difficult time to come to terms with such experiences. The weirdness of such description has remarkable resemblance to the weirdness of the quantum world. In fact, one would need a universal quantum computer model of one grand mind to explain these. For example, as Buddha was able to know and affect anything through mere intentionality irrespective of distance. It almost appears that he had access to grand nonlocal quantum computing core and was able to access and affect any information at will. Kabir described his experiences of this inner world as a reality where opposites existed together in one state. For example, he describes fire and water coexisting as one state. Such dualities can only coexist in a qubit, but ancients had no inkling about quantum BITs or computer. They simply documented what they saw. Numerous others, throughout the history of human kind, have experienced varied degree of experiences of entering into this informational space. In her book Fingerprints of God: The Search for the Science of Spirituality, National Public Radio correspondent Barbara Bradley Hagerty provides an in-depth account of many of such experiences.

The word "dhayan" in Hindi language is synonymous with the process of meditation. Dhayan actually means attention. The quality of attention is very critical. The concentration of your attention is equally important. What your attention is focused on is particularly critical. Ancient seers found "the

It is easy to say that God in its totality is everywhere but the true magic occurs when I understand that he in his totality is inside of *me* like quantum hologram, complete hologram is part of all points on the hologram. He is in information regime, nonlocal regime, beyond space and time.

—Anonymous

will" to be the most noteworthy affecter of their attention. Your attention will act to fulfill your intentionality through your conscious will. Your intention or subconscious, therefore, plays a crucial role in directing your attention. When your conscious will is not controlling the attention, it is the intention living in your subconscious mind that takes over. It uses the importance you assign to a particular process or event as its guide. It almost works as if there is a list, which is sorted in order of importance attached by you. Understanding this importance is one of the most critical components of achieving enlightenment. The word "gyan" in Hindi translates into English as knowledge. It means a particular type of knowledge that affects or can affect the importance of various tasks on the list that intention works with. So the knowledge that affects your list of priorities to the point that communion with divine becomes the top of your list is considered the true knowledge.

Today there is a lot of the scientific basis and recognition that these ancient processes are capable of altering of the human state of mind in a significant way. Several scientific studies have shown that meditation Samadhi or other similar spiritual practices can put human mind into what they described as an alternate state of reality. Similar effects are reported for individuals who believe that they have achieved communion with God through intense prayers. In her book Fingerprints of God: The Search for the Science of Spirituality, Barbara illustrates numerous scientific studies that have examined the divinelike alternate realities induced by spiritual experiences and their remarkable similarities with chemically induced experiences of divinity. She describes the individual who are capable of achieving such state at will as spiritual virtuosos. Just like star athletes, musicians, or CEOs, the understanding, the knowledge, and the skills to be one must be mastered, she believes. Even though the altered state of reality that leads to God experience is undoubtedly not a trivial state of mind to achieve by most humans, but it is the most desirable state for most humans to be in. Why is so? It is because it is the most pleasurable state of being bliss sprouts from the core of your being. It is the promise of enlightenment. It is the supreme experience, which our brains are wired for. According to ancient Ayurvedic seers, all humans with appropriate knowing or self-knowledge can achieve that state.

Ancient seers were masters of human psychology. They studied human brain intensively, however, mainly from subjective view. Just like modern psychologists, they also realized that the human mind processes information

Two ways to commune with your nonlocal mind or quantum self.

1. Silence between the gaps of your thoughts
2. Remembering and accepting its true nature

—Anonymous

in four fundamental ways, thinking, feeling, sensing, and intuitive. All four of these abilities of the human mind help to form a reality of our outer and inner world. They realized that out of the four abilities of the human brain, the most fundamental was the feeling. Without feelings, life could not be defined. Today, there is a lot of the scientific basis for such a realization. Our emotional brain is spread throughout our body forming a highly fluid informational network of molecules.

Each feeling that we create reflects a string of complex information pattern represented by the chain of biochemical reactions in our body. When we feel positive, our mind/body produces informational molecules that improve our inner healing intelligence and stimulate our immune system. Every thought we think, every emotion we feel is packaged in informational molecules that spread the message throughout our body. All these molecules of emotions are finally rooted in quantum information processing just as emotional mind is eventually rooted in the cosmic mind or God.

According to Pert, author of the book Molecules of Emotion, "at a neurological level, the feeling of being connected with God, of feeling blessed, is an important part of the brain. Blessing and bliss come from the same root. We are hard wired to be in bliss. It is normal, and it is natural. There is a straight evolutionary argument for this function—any creature that could not experience bliss would have just died and become extinct 200 million years ago. It is as we are designed to make choices around pleasure. The very highest, most intelligent part of our brain is drenched in receptors to make us use pleasure as a criterion for our decisions. So it is okay to feel good—God is good."

The ancient Ayurvedic seekers knew extensively about the bliss response and these receptors in our bodies. Their texts describe emotional energy centers called chakras, which are located in different areas of the body. The location of these chakras matches well with areas showing high concentrations of informational molecules in our bodies. The vibrating informational molecules are the gateway to divine the supreme quantum computer operating this universe.

The heart chakra is located in our heart. It is also referred to as the fourth chakra. In Sanskrit, it is called the anahata chakra. It is said to be the

> State of matter comprising of human particle, in living state of love, gratitude and acceptance but silent like dead . . . Samadhi
>
> —Anonymous

emotional energy center that enables you to feel higher emotions, such as love, compassion, forgiveness, tolerance, happiness, and joy. It is also the basis of widely held belief, that, it is your heart, which allows you to feel divinity. Heart is the center of your emotions and the home of your higher consciousness. Activating and balancing the heart chakra expands your consciousness, from love to devotion, and to ultimate surrender. Love, devotion, and surrender are like the three rungs in the spiritual ladder. The first rung is love, the second is devotion, and the final is surrender. Love is sweet, devotion is sweeter, and surrender is sweetest.

According to Ayurvedic seekers, we love because at every moment we hunger to realize the highest, to feel and to be consciously one with the whole. Human love binds and is bound. Divine love expands and enlarges itself. India's greatest poet, Rabindranath Tagore, said, "We approach God. He is our dearest, not because He is Omniscient and Omnipotent, but just because He is all Love." Devotion is the intensity in love, and surrender is the fulfillment of love. Surrender is protection, and surrender is illumination. Surrender is dissolution of the binary self. Surrender is our perfection. A tiny drop enters the ocean and becomes the mighty, the boundless ocean.

We often feel that if we surrender to someone, which means we accept his or her power over us. From egoic entity point of view, this is true. From the quantum self point of view, it is incorrect. When the finite enters the infinite, it becomes the infinite all at once just as a tiny drop enters the ocean and the drop cannot be traced. It becomes the ocean.

"Everyone knows there are many drops in the ocean, but few know there is the ocean in the drop. I want you to be the ones who know the ocean is in the drop," says Prem Rawat or Maharajji to his followers.

Kundalini is the Sanskrit term used to describe the phenomena of awakening or enlightenment. The Christians saints often describe it as the fiery descent of the Holy Spirit. The term describes dormant energy that begins to rise or release through the body. It opens the chakras or releases stored and blocked emotional energies. It almost feels as if a snake like entity is rising within the body. It can be quite intense and often leads to mental states that seem beyond any experiences of this world.

My soul likes the *truth*. There is no bigger *truth* than *now*. When my experiential reality (feeling) encounters the real *me*, it rejoices.

—Anonymous

What awakens the Kundalini? The Kundalini awakening can be caused by many circumstances such as extended periods of meditation, yoga, or fasting. It is intriguing that in some cases, stress, trauma, psychedelic drugs, or near-death experiences may induce similar experiences. After centuries of subjective experimentation of sensing, feeling, and thinking brain through introspection, ritualistic, prayer and meditative methods, ancients discovered this extraordinary emotional state of being, the awakening of Kundalini.

Ancients described the process of awakening Kundalini as the motion of fire serpent within one's body. When not awakened or under normal conditions, it sleeps in a state coiled three-and-one-half times at the base of the spine at the root chakra. As the awakening begins, it uncoils inside oneself, releasing a swarm of emotional energy as it goes and opening the chakras, until its fiery head pokes out through the crown chakra, signifying enlightenment. When one experiences a Kundalini awakening, it changes their life forever. Everything will have the flavor of an enlightened consciousness behind it.

In an awakened Kundalini state, one may be filled with the most incredible states of blissful cosmic energy. This is an experience of unlimited comprehension of reality, a taste of the greatest healing energy available inside and feelings of absolute ecstasy. After an experience of Kundalini awakening, true nature of reality will be clearly visible to you. You will experience constant connection to higher intelligence everywhere with the abundance of love and healing. You will see the world as sacred and divine as it has always meant to be. According to ancient seers, this is what true enlightenment is all about!

Road to Divinity

Dissolution of Egoic Entity, Embracing Quantum Thinking

The first and perhaps the most difficult step for modern humans in achieving enlightenment is the dissolution of the egoic entity, which resides in all of us. Egoic entity is the gift of modern living. It is the largest benefactor of binary thinking. This is the thinking that Buddha called as dualistic thinking and held it responsible for all dukha or sufferings. Most of us support this entity residing in our deep psyche. Some have more, and some have less. Spiritually speaking, the mass is equivalent of the ego bounded in binary thoughts,

My egoic self exists because of dissatisfaction. Content self is selfless. It is ordinary and abundantly available in nature. Ego does not accept the ordinary. It is never happy with what it has and always wants more. Ego learns to be perpetually dissatisfied. Until it is no more.

—Anonymous

emotions, convictions, and beliefs of the mind. The more frozen are these man-made egoic convictions, the bigger is the ego. Enlightenment is then the process of dissolving the ego to transform it to the pure consciousness or true free will or true quantum self. From this point of view, one may explain en-lighten-ment to literally mean the process of becoming lighter or less burdensome or lesser mass.

An ego does not dissolve itself. In fact, our modern society constantly reinforces it. Doubt is at the foundation of this entity. Doubt is truly the lack of trust. You do not trust because you are fearful. You must transcend your fears. Doubt is not bad by itself, but it is not the end. One needs to play doubt fully and go beyond. Egoic entities form egoic networks. In this network, fear and greed dominates. Recognition of self in this network is through wealth you have, the position you occupy in the society, resources you control, type of house or houses you have, and type of automobile you drive. Egoic thinkers are often fearful and in a fight or flight mode. They often end up seeking to control where the control is not possible. Freedom simply translates into taking most they can from the system so they never have to depend on the network again. Financial freedom is highly desired achievement among the members of egoic network.

Since most humans are part of the human society and depend on these egoic networks for food and shelter, they are all part of this network in small or large ways. Pulling out of the egoic network is quite a task in itself. This is the biggest roadblock in taking a step toward dissolution of egoic entities. The egoic network is the network of endless dissatisfaction. Even at its peak level of achievements, it fails to provide the satisfaction, contentment, and true peace of mind to members who were able to achieve such successes. Only a few get to experience this stark reality. The network is propagated through recruiting new members primarily with the lure of wealth and power. Therefore, there is an implicit need to ensure that most members do not know the limitations and shallowness attached to be a part of this network. Most spend their lives being part of the network and suffer until death. In most cases, family members are all part of the network, and it becomes harder and harder for them to identify any existence outside the network.

Members of this egoic network are in constant fight with each other for mega portion of material wealth and power. This leads to a constant struggle and

> *What do wise say about tomorrow?*
> If it comes fine, and if does not come fine . . .
> Please don't bother me . . .
> Can't you see I'm busy . . .
> Enjoying *now*.
>
> —Anonymous

lack of mental peace. Chronic conditions of stress, along with the ailments resultant from chronic stress conditions cause some members to find a way out. Many come across various religious groups, spiritually enlightened masters, etc.; all of them attempting to steer them away from the egoic thinking and traumatic effects of their egoic entity on their inner world.

Understanding is the key ingredient of dissolving egoic being residing in us and activating quantum thinking. Many resort to charity or volunteer work, which is the cornerstone for quantum thinking. Understanding one's true nature of quantum self is essential. Slowly all the doubts dissolve, judgment of others gets eliminated and is replaced by trust. A powerful understanding of all, as part of one, emerges. Quantum thinking is primarily thinking for the benefit of others rather than the benefit of the binary self. The thinking arises out of paying attention to messages received from the cosmic mind through emotional mind. Since our body evolved from a quantum system, the breathing, heartbeat, and other involuntary processes are rooted in the cosmic, or quantum thinking. An egoic mind can shift its attention from the binary thinking to the quantum thinking by simply paying attention to these involuntary processes. Such practices are widely followed by most holy masters under the name of mediation, Samadhi, prayers, etc. Later I will describe a breathing technique of self-observation that is highly effective in dissolving egoic entity.

The Power of Present Moment, Your Attention

A second essential step to realize enlightenment is learning to focus your attention, especially at the present moment. This is a powerful and practical way to dissolve the influence of your egoic entity on your emotional state of being. It also means that you have chosen to be with your quantum self over the egoic entity that also resides within you. All the past moments came to us in present, and all the future moments will also come as present. Our conscious mind is our attention, which resides in present. Mastering the attention was the key skills that ancient Ayurvedic seers honed to achieve the state of Samadhi. Our intentions or one may call as subconscious mind is stuck in the past or future in the form of thinking. The present is truly a thoughtless state. Egoic entity wants you to stay in the past or the future, but never in the present. Therefore, most people who are part of the egoic network exist to serve their egos by ignoring the present and clinging on to either past or future. A binary

I without mine is,
My soul,
My pure consciousness,
The real me.

—Anonymous

thinker is never close to the truth as rational mind can only go so far to grasp the whole truth. Eternal truth only resides in present. A quantum thinker acknowledges the massive speed and power of parallel quantum processors at the core his true being and simply accepts the inputs in form of incredibly delicate and soft feelings or waves of guidance from it.

Once you learn how to live in the present, your whole energy, your whole purpose goes into the present moment. This means you put everything you have into the present action and that action becomes perfect basis for the perfect future.

If you take time to live in the present and observe, touch, smell, and feel what is around you, you put your whole mind at rest and start seeing true colors of life. Your ego does not have anything to feed on since you are not using your binary mind to worry, plan, or speculate about the events to come. Only awareness and acceptance is left. It guides you to achieve effortlessly everything that you desire.

Knowing Your True Self

The third step of achieving enlightenment is knowing the nature of your true self. I go back to the discussion I initiated in the first chapter of this book about the self that needs to feel good. Who is it and where it is? In some sense the key message from this entire book has been to understand the nature of this self.

We started our journey with the first book of the series, Road to Digital Divine. We have realized that our universe is highly computational with information at the root of its being. A simple picture of my universe and me has emerged. From a top-down view, it has become clear that everything I see in physical space-time universe is reducible to information comprising of a string of simple quantum or binary decision-making elements. All that exists in our universe and is visible can be reduced to informational entities comprising of these strings. All quantum particles, all atoms, molecules, cells, all matter, all energy, all living beings including humans are informational devices with linked strings of elementary quantum or binary decision makers. These links have given rise to simple or complex informational entities that are capable of processing information.

Divinity exists as part of *your true* reality, inside us, gateway to the rest, you can interact with it like a real person, you can reduce your discontentment, anxiety, and anger by handing over your problems to *him*, *you* can rejoice express gratitude and appreciation for all *he* does for you.

—Anonymous

Information processing is what my universe and all its inhabitants do all the time. In fact, this may be the most basic nature and pretty much the only action that goes on in our universe. All of them observe, send receive, and modify informational signals locally as well as nonlocally to each other. An army of observers reduce quantum information strings to binary informational strings. Some entities perform computation according to the rules of quantum information processing while others follow the rules of binary computation. Life in general is rooted in both the systems of computation. It is computationally astute enough to take advantage of its quantum roots as well as translates that to a binary processing that can operate its large, classical body.

It has become clear that myself is more than what I see in the mirror as my physical body. The most appropriate way to understand myself is through informational, computational account of my being. My true self is interplay of three minds that co-ordinate and manage all information exchanges. These are, the cosmic mind, my emotional mind that is my body, and my rational mind which resides in my emotional mind.

I find my physical, human body living in two universes. One is eternal, where the wave of pure potentialities and universal mega quantum computing resides. It is the timeless, spaceless and energyless dimension where everything is eternal. Whatever was, is, and will be. The second is the physical universe of space-time where my visible, physical body exists. It is subject to entropic decay and subsequently dies like other entities in this space-time universe. My senses and my abilities to reason keep me aware of what is going on in this world of physical devices.

My emotional mind resides at the union or interface of these two universes. While my body is aware of its needs arising out of its living in the physical domain, it is also aware of the influence of the eternal domain by feeling the waves of potentialities, mathematical completeness in the eternal universe. My true quantum computational self resides in this dimension as a superpositional state of my being. Here it is always in complete contact with the rest of the universe and considers all to be part of its self. My egoic self resides in space-time dimension. It thinks it is separate from the rest and defends its existence through acts for the self. When my binary or rational mind dissolves its confining conceptual boundaries, it merges into the quantum or nonlocal one wholesome consciousness, it becomes whole for eternity.

When I surrender or merge and ride with that feeling
with full acceptance, humility, and devotion,
When I'm not distracted by unconsciousness (thinking)
My journey to timeless realm begins . . .
It is the divinity that I have always craved for . . .

—Anonymous

My binary self is a master of binary computation where as my quantum self is what holds all my quantum decision makers together. This is the last stop for my top-down perspective. It is also the last stop for doubt or reason, which has been my companion to know material reality. But now to have a deeper bottom-up view, I must transcend the doubt or reason. For this, the binary entity must dissolve into the quantum self. It must transcends the physical world of space-time and enter into the eternal world of quantum self. I must know and embrace my true nature, the true nature of my quantum self. My rational mind must transcend reason or logic to accept super reasoning that binds all in unified oneness. I must understand that my true nature is not of doubt but of trust and love.

Just like my quantum self exists in this eternal regime, everything I perceive in the physical world has an informational equivalent in this eternal universe. Therefore, my parents, my son, my friends, my colleagues all are computational entities in the eternal universe of my consciousness. My awareness provides me with complete map of the computational universe residing inside of me, according to my perceptual reality. The way I look at them in my internal world affects the way I look at me in the physical world. If I am being nice or not nice or hiding something from someone in my physical world has an impact on my inner world through these computational entities and vice versa. My interaction with all entities is not limited at external space-time level but also extends to internal which is in informational regime. Ultimately, emotional state of myself is affected by what goes on not only in my external world but also in my informational world or inner world.

Myself, in the eternal regime, needs to be in its true state of being. It is a collective state of being. This is the state of utter contentment or mathematical completeness or self-satisfaction or complete harmony or equilibrium with the rest, completely trusting of the whole, being in the present moment, silent witness, and truthful. The ancient seers recognized these needs of the true self, and its effect on emotional state of being, therefore, felt the need for the physical being to be content, empathetic, harmonious, and truthful. This is the central teaching of yogic science.

Enlightenment is knowing the ultimate truth. What is that truth? The truth is the true nature of us, the true nature of the universe. True nature of how we are all connected. The truth that we are all truly one not separate entity.

Your attention is like monkey. It moves from one place to another. When it merges or communes with heart and accepts (with humility, gratitude, appreciation, and devotion) Param chit anand (supreme bliss) results.

—Anonymous

That there is a part of us that never dies or is eternal, and there is a part of us, that dies like our physical body. That there is a quantum computational divine core that binds us all together. The pain and suffering others feel are genuinely pain and suffering we feel too.

Once this truth is understood from a mental perspective, there is one more step left to achieving true enlightenment. It is to experientially feel and dance with the divinity. Let the discussion that follows be your guide. Practice the techniques at least once a day for as long as you can. See if you can make a goal of giving yourself a gift of one hour per day of quite time, just to learn true nature of yourself. Longer you practice; more positive the impact, these techniques will have on your emotional state. Many of our modern diseases are direct results of our paying too much attention to binary thinking. As soon as you learn to see beyond this type of thinking, you will begin to experience relief from those conditions. I was personally able to reduce significantly my dependence on medicines following these techniques. How do I know that I am dancing with the divinity? "It will be recognizable through the fruits of your efforts," says Maharajji. When it will happen, you would know; in fact, every particle of your being will know. It will fill your heart and mind with incredible joy pouring out of your deep inner being.

How to Dance with Divinity and Reach Enlightenment

Even though, the Ayurvedic seers of the past never had a clue about the computational nature of the universe and of the human body, through their introspection and self-experimentations, they arrived at strikingly close conclusions. The dissolution of binary self into the informational whole and witnessing such transition with the quantum self results into this incredible knowing or feeling of dancing with divinity. They understood that being in the present time and attending to the nonlocal mind through emotional mind was the key to be in bliss state. This wisdom can benefit us all. I would extract the core of their teachings and explain in our modern world language so all can understand. Here is how all humans can reach this extraordinary state of being.

Step 1: Be in the present state of awareness via self-observation.

At any moment, our mind is involved in thinking a number of activities. This is not being in the present state. The rational thinking mind must be

> I can reach that place by flying on a feeling; to fly I need to reduce my weight by being conscious (no thinking) or surrender to that feeling.
>
> —Anonymous

disengaged to be in the present state. We can understand this by a simple analogy. Consider our rational thinking mind as the personal desktop computer. This is not particularly far-fetched. As we have seen, our rational mind is actually a binary computer and follows a remarkably similar computation scheme as a personal computer.

We know very well how to open a program and close a program on our desktop computer. However, do we know how to open and close programs running in our thinking rational mind?

We know very well that we cannot keep a lot of programs open on our desktop computer as the speed of computation gets affected once too many programs are open at the same time. Do we know that a similar analogy exists when we take on too many thinking tasks and our rational mind cannot keep up with all that comes its way?

We know how to open a program in our thinking mind. Any simple act of thinking, memorizing, seeing, or doing opens a new file. However, do we know how to close them as well? In fact, most of us are not careful and run into circumstances where we keep opening these programs or data files in our thinking mind. Our thinking mind, like a high fidelity computer, gets engaged in tackling these issues. We eventually end up opening too many programs running in our thinking mind that our systems get overwhelmed, and we begin to feel miserable.

We all know that the effect of our thoughts is not limited to our rational mind. Out rational mind recruits our emotional mind to engage the whole being to attend to the subject of our thoughts. Therefore, the files are opened by our thinking mind, stay open in our emotional mind, which runs throughout our body. The thinking mind usually moves on to solve the next urgent item on the list; our emotional intelligence follows it like a high fidelity server and keeps opening more and more files. Health of our emotional mind is predicated on the proper management of information flowing through the emotional brain. In particular, the critical are the avoidance of dualistic thoughts specially geared to maximize ones self-interest (egoic entity) and not in the interest of all. Ancient seekers called these thoughts as impurities. Ancient seers developed techniques to close such files or programs.

A mind immersed in silence is the mind of a child. The curiosity, the love, and appreciation are the inherent properties of this mind.

—Anonymous

Ancient seekers discovered that whenever any such thinking impurity arises in mind, the breath loses its normal rhythm, and we start breathing harder. One can observe this. Thoughts, emotions, and sensations are linked trough informational circuitry. The molecules of emotion that are floating in your body making this incredible information link possible.

One may use this information link in a practical ways to stay aware or in tune with one's thought process or emotional state of being. Any normal person can be trained to observe respiration and body sensations, both of which are directly related to informational patterns of information in our mind caused by change in our emotional states such as fear, anger, love, or passion. Through a practice of observing our respiration or sensations in our body, we are, in actuality, observing our own mental impurities.

Following our earlier analogy of a desktop computer, impure thoughts open the files of negative emotions in us. By observing and accepting the feelings, we are effectively able to put a closure to these open files. As a result, we discover that these negative emotions lose their strength. If we persevere, they eventually disappear as all these negative emotions containing files close. We begin to achieve a peaceful and happy state of living, a life which is increasingly free of negativities.

In his book Breath Sweeps Mind, Zen Master Jakusho Kwong describes in details the deep meaning behind this title phrase, which came to his consciousness when he was just a young man in his twenties. He wrote this phrase on a rice paper but did not truly understand the significance of this phrase. According to him, only after decades of practice he truly understood the true meaning of the phrase as he was able to sweep negative thoughts out of his mind with each breath. Prem Rawat, a well-recognized master, says, "Only if one can understand the meaning of this coming and going of breath, one can understand the meaning of it all."

Step 2: Recognize, accept, and surrender to the wave of guidance.

The information wave of potentialities guides all entities, small or large. It is just as the photon we saw earlier was guided by this wave. These waves are instantaneous and bring with them a message from the grand mind. The recognition and acceptance is a start. Complete surrender to it and

True contentment means no reason to be engaged in action, unconscious (subconscious) thought, or programming. Complete surrender to the divinity with feeling of appreciation and gratitude.

—Anonymous

utterly hypnotized by its presence and in complete love with it represents the state of enlightenment. The more one understands this and practices this approach and accepts the guidance, the more quickly thought impurities would dissolve. Gradually, the mind becomes free of the binary self and becomes pure. Our true quantum self takes over. A pure mind powered by the cosmic mind is always full of love. It is selfless love that promotes existence for others. It is a mind full of compassion for the sufferings or failings of others. It is a heart that rejoices at others successes and achievements.

As one reaches this point, the mind is balanced or in perfect harmony. A balanced mind not only becomes peaceful but also permeates peace in the surrounding atmosphere. One becomes sensitive to the sufferings of others, doing whatever they can in order to provide help with an attitude full of love, compassion, and calmness.

With greater acceptance and surrender to feelings arising out of pure mind, true nature of self begins to reveal itself. What is needed, then, is to "know thyself," not just intellectually but experientially. As one deepens the practice, one may see the ultimate truth of matter and mind. As one transcends beyond the confinements of perceptual calibration, beyond time and space, the truth of total liberation and love is experienced. It is the ultimate goal of everyone.

What does this experience feel like? This is the most frequent questions asked by most. There is no answer that can be given which is complete and justify the true intent of the question. This is because the answer and the description will come from a binary processor, and it is truly an experience of quantum processor. There is a risk involved here as well. A description may create expectations, which may become the barrier to get there. You must go there to experience it for yourself. Yet, our modern upbringing forces us to identify characteristics before we spend our time and energy exploring it. It is the call of the egoic entity within us. Only a general guide to this state of being can be provided. There are three main feelings described by many. First is the extreme feeling of contentment, joy, or bliss that appears simply to pour out of one's inner being. Second is clarity. Everything appears to be in the right place and at the right time. Third is the experience beyond space and time or concepts. Time appears to stop, or one may feel timeless or eternal.

At the deepest level of my existence there is
this profound oneness, profound fullness or
contentment or completeness. The witness is
lost in the rest. Eternity has arrived. The feeling
of immortality, timelessness, spacelessness,
weightlessness become the nature.

—Anonymous

Enlightenment is ultimately a subjective process, a process of knowing the ultimate truth. The ultimate truth is God. God is realized in the form of an all-encompassing feeling that compels the receiver in a state of truth, total acceptance, pure love, compassion, timelessness, and oneness. All doubts disappear; gratitude and appreciation reign supreme. May all experience this ultimate truth. May all be free from sufferings. May all enjoy complete peace, real harmony, and true joy from deep within.

———◦◦◦◦◦◦◦◦———

For more Information and free downloads, visit

http://www.drhemantgupta.com
or
http://www.roadtodigitaldivine.com

Notes

Chapter 1

As I was researching the way to introduce self and introduce the meaning of "I" we all take for granted in our expression I came across this on the Internet:

Expression of Self, Templeton-Cambridge | Emily Yoffe—[October 16, 2008], http://www.templeton-cambridge.org/fellows/showarticle. php?article=176.

> *Original Poetry,* "**I Know Not Who I Am**", *Shishir Gupta: Write. Learn. Share. http://www.originalpoetry.com/poet/ecrit. The inspiration and command to write this book may be divine. It, however, manifested in highly nonlinear manner. Before I started writing this book, I had no aspiration to be a writer. A series of apparently disjointed events initiated me on this path and stayed with me all the way to completion. It initiated with Shishir Gupta, my younger brother. Shishir is a talented soul, a banker by training. His intellectual depth is quite unappreciated by his work environment as he has been struggling to find an outlet to flow his talents. Finally, I think, he has found his calling. It is poetry. It allows him to express what resides deep within his being. A few years back he sent me this poem. His intention was for me to help this incredible poetry to simply be published. Having an emotional depth to appreciate such a creation was neither needed nor expected of a scientific, utterly rational creature, such as myself. Ordinarily, I would not be much influenced by such nonscientific literature, but the timing was such that when I first read it, my internal struggle with the concept of self was at paramount. The first reading stirred something very deep within. As I read this marvelous creation, again and again, what was inside of me began to melt and started taking shape of this book. I am very thankful to Shishir.*

Road to Digital Divinity, The first book of this series Informational nature of Being, The nature and origin of three minds, viz. rational, emotional and Cosmic are described in details in first book of this series.

Chapter 2

"The Wysiwyg Universe." http://razorland55.free.fr/Word/wysiwyg.pdf.

"Does Time Exist?": IntentBlog, http://www.intentblog.com/archives/2008/01/does_time_exist.html

"Quantum Information Theory Detail "—Perimeter Institute for Theoretical Physics http://www.perimeterinstitute.ca/Outreach/Explore_Our_Universe/Quantum_Information_Theory_Detail/.

"Queueing systems—CiteSeerX citation query," http://citeseerx.ksu.edu.sa/showciting?cid=4161.

"Did Einstein Claim That Nature Has Mathematical Structure?", Jaroslaw Mrozek, http://itis.volta.alessandria.it/episteme/ep6/ep6-mrozek.htm.

Aurablog.org · "Working with the Light." http://aurablog.org/category/working-with-the-light.

"Biological Utilization of Quantum Nonlocality." http://www.meta-religion.com/Physics/Consciousness/biological_utilization_of_quantu.htm.

Quantum Approaches to Consciousness Stanford Encyclopedia article. http://www.quantumconsciousness.org/StanfordEncyclopediaarticle.htm.

"Imaginary Numbers are not Real—the Geometric Algebra of." http://www.mrao.cam.ac.uk/~clifford/introduction/intro/intro.html.

Shift in Action. *"The Biology of Belief"* http://www.shiftinaction.com/node/2464

Not PC: 2006-03-12, http://pc.blogspot.com/2006_03_12_archive.html.

"Seattle Personal Injury and Car Accident Attorney: The Tragic," http://www.facebook.com/note.php?note_id=60053889699&ref=mf

"Sleep Talking: Sleep as an Altered State of Consciousness," http://www.open-spaces.com/article-v7n4-sack.php. http://www.merrynjose.com/artman/publish/printer_118.shtml.

"Ultimate Computing," http://www.gaianxaos.com/pdf/consciousness/ultimate_computing.pdf

Danah Zohar, The Quantum Self, Arizona > CONSCIOUSN 2008 > register (2009-02-06 20:25:31), http://www.coursehero.com/file/1370389/register

Nick Herbert, "Consciousness and Quantum Reality," http://twm.co.nz/herbert.htm.

Chapter 3

Calif. robber confesses to priest, turns in cash, uhttp://www.msnbc.msn.com/id/32182102/ns/us_news-weird_news/?GT1=43001

"PLEKTIX: Altruistic and Selfish Bacteria," http://plektix.fieldofscience.com/2008/04/altruistic-and-selfish-bacteria-coexist.html.

Chapter 4

"The Origins of Selves," http://ase.tufts.edu/cogstud/papers/originss.htm (accessed February 17, 2010).

Mix ID 5882, http://www.rssmix.com/u/5882.

"Neuroeconomics and Social Decision-Making: The Frontal Cortex," http://scienceblogs.com/cortex/2007/11/neuroeconomics_and_social_deci.php

Herbert Simon: Biography from Answers.com, http://www.answers.com/topic/herbert-simon.

"Neuroeconomics and Social Decision-Making: The Frontal Cortex," http://scienceblogs.com/cortex/2007/11/neuroeconomics_and_social_deci.php.

"Bacteria Offer Insights into Human Decision Making," http://www.physorg.com/news179521562.html.

A-One Home Shield (AHS) is not a real organization as far as author's understanding goes. It is a disguised name for the actual organization involved in the incident.

A Beautiful Mind, John Nash, http://mockingbird.creighton.edu/english/ fajardo/teaching/srp435/beautiful-mind.htm.

"Hug the Monkey: Autism Gene Linked to Empathy," http://www. hugthemonkey.com/2009/11/autism-gene-linked-to-empathy.html?no_ prefetch=1. presentation-on-legal-knowledge-management-by-kwami-ahiabenu, http://www.scribd.com/doc/2680469/presentationonlegal knowledgemanagement bykwamiahiabenuii 200711944917948614402.

Indonesia Software Development Services Company, http://www.iaodt. com/aboutus-credoVision.aspx.

The origin of DIKW Hierarchy, http://nsharma.people.si.umich.edu/ dikw_origin.htm.

"The Origins of Selves," http://ase.tufts.edu/cogstud/papers/originss.htm (accessed February 17, 2010).

Mix ID 5882, http://www.rssmix.com/u/5882.

"Neuroeconomics and Social Decision-Making: The Frontal Cortex," http:// scienceblogs.com/cortex/2007/11/neuroeconomics_and_social_deci.php.

A Beautiful Mind, John Nash, http://mockingbird.creighton.edu/english/ fajardo/teaching/srp435/beautiful-mind.htm.

"Hug the Monkey: Autism Gene Linked to Empathy," http://www. hugthemonkey.com/2009/11/autism-gene-linked-to-empathy.html?no_ prefetch=1. presentation-on-legal-knowledge-management-by-kwami-ahiabenu, http://www.scribd.com/doc/2680469/presentationonlegal knowledgemanagement bykwamiahiabenuii 200711944917948614402.

Valeria Fugante's Blog, Blog Archive, "From Data to Wisdom," http://blogs. southworks.net/vfugante/2008/09/06/from-data-to-wisdom/.

Confession of an Ordained Minister, Sujata C., "The Final Test by," http://www.bolokids.com/2008/0639.htm.

"Evolution of Altruism," http://serendip.brynmawr.edu/sci_cult/evolit/s05/web2/amnuskin.html.

Story of Selfless Heroism, Dan Abrams, http://www.jewishworldreview.com/0705/abrams072605.php3?printer_friendly

Complete text can be found: Being a Gatekeeper | Merrill Associates, http://merrillassociates.com/topic/2004/04/being-gatekeeper.

LOVE. Gary Amirault, founder of Tentmaker Ministries.

Professor Braj Kishor, a retired professor of business management from Osmania University, India, has been my soul force in developing this book from start to finish. My weekly telecom with him for last several years, his daily emails to me, his son, of his self-accrued as well as eternal wisdom guided the content all through.

Mohandas Karamchand Gandhi Biography | World of Sociology, http://www.bookrags.com/biography/mohandas-karamchand-gandhi-soc

Prem Rawat or Maharajji http://www.tprf.org/

Osho (Bhagwan Shree Rajneesh) www.osho.com

Candace Beebe Pert, Molecules of Emotion: The Science between Mind-Body Medicine, Scribner (1999).

Candace Beebe Pert, Everything You Need to Know to Feel Go(o)d, with Nancy Marriott, Hay House, Inc. (2006).

Deepak Chopra, 1988, Return of the Rishi.

Deepak Chopra, 1989, Quantum Healing: Exploring the Frontiers of Mind/Body Medicine.

Deepak Chopra, 1997, The Path to Love: Renewing the Power of Spirit in Your Life.

Deepak Chopra, 1997, The Seven Spiritual Laws for Parents: Guiding Your Children to Success and Fulfillment.

Deepak Chopra, 1999, Everyday Immortality: A Concise Course in Spiritual Transformation.

Deepak Chopra, 2000, How to Know God The Soul's Journey into the Mystery of Mysteries.

Deepak Chopra, 2003, The Spontaneous Fulfillment of Desire: Harnessing the Infinite Power of Coincidence.

Deepak Chopra, 2003, Synchrodestiny: Harnessing the Infinite Power of Coincidence to Create Miracles.

Deepak Chopra, 2006, Life after Death: The Burden of Proof.

Deepak Chopra, 2006, Kama Sutra: Including the Seven Spiritual Laws of Love.

Deepak Chopra, 2007, Buddha: A Story of Enlightenment.

Deepak Chopra, 2008, Jesus: A Story of Enlightenment.

Chapter 5

TED Blog: Henry Markram at TEDGlobal 2009: Running notes from, http://blog.ted.com/2009/07/henry_markram_a.php.

"Has Science Found God in Nonlocal Reality?" http://www.plim.org/nonlocal.htm.

"Hinduism & Quantum Physics," http://www.hinduism.co.za/hinduism_quantum.htm.

"The Glory of God," http://www.seekgod.org/message/gloryofgod.html.

Chapter 7 of Srimad BHagwat Geeta (Chapter 7), http://sites.srimad bhagwatgeeta.com/chapter-7/.

KABIR POEMS, http://midiagols.com.br/africa2010/quadro.php? help=Kabir+Poems.

Fingerprints of God: The Search for the Science of Spirituality, National Public Radio correspondent, Barbara Bradley Hagerty.

Love, Devotion and Surrender—Sri Chinmoy, http://www.srichinmoy.org/ polski/resources/library/talks/inner_qualities/love_devotion/.

BBC NEWS | Science & Environment | Simulated brain closer to, http:// news.bbc.co.uk/2/hi/science/nature/8012496.stm.

"How to Awaken Your Kundalini" http://www.enlightenedbeings.com/ how-to-awaken-kundalini.html.

"Biology of Kundalini—Biological Relation to Zero-Point Energy," http:// biologyofkundalini.com/article.php?story=BiologicalRelationtoZer o-Point.

"Vipassana Meditation: Art of Living," http://www.pl.dhamma.org/index. php?id=1366&L=0.

Throughout writing of this book, various quotes simply came and stayed in my mind and I was able to write them down. These are the Quotes designated from Anonymous as I do not know the origin of these.

All images used in this book have been obtained from Wikimedia Commons which in turn cites images source as Flicker's The Commons.

About the Author

Hemant Gupta is the author of Road to Digital Divine, first book in the series Informational Nature of Being. This series of books reflect his quest to understand true nature of himself and his universe. He graduated from University of Southern California, completing his masters and PhD degrees in polymer science. He subsequently finished an executive MBA course from University of California, Los Angeles. He has been vice president of research and development at a major corporation for the past ten years and in various leadership roles for the past twenty years.

Index

www.ingramcontent.com/pod-product-compliance
Lightning Source LLC
Chambersburg PA
CBHW032007170526
45157CB00002B/583